ROLEX GENEVE
17 RUBIS

OSVALDO PATRIZZI and MARA CAPPELLETTI

ROLEX

HISTORY, ICONS AND RECORD-BREAKING MODELS

ACC ART BOOKS

CONTENTS

DETAILS OF STYLE

THE LOGO ON THE WINDING CROWN

The origin of the term "Rolex" is unknown. While Hans Wilsdorf never provided any confirmation, according to some people it comes from the French expression "horlogerie exquise", or "exquisite watchmaking". According to others, it is simply a name that was chosen because it is easy to pronounce in all languages. The distinctive five-pointed crown logo, the long-established symbol of the *maison*, was introduced in 1925.
Since then it has featured on the winding crown of almost all models.

THE HANDS

For the parts of the Professional collections, Rolex generally employs Squelette or hollowed (Mercedes) hands.
For the Oyster Perpetual series, it uses baton hands (Parformes) instead. All are treated with luminescent 'Superluminova', for visibility in the dark.
The hands are usually made of the same material as the case; if this is made of gold, the colour is also the same as that of the ring or bezel.

THE REHAUT

Since 2004 almost all Rolex models have featured a flared ring called rehaut (raised section) that is part of the carrure and is located between the glass and the dial. Engraved on the side of this raised section, at 12 o'clock, is the logo, along with the brand – repeated a number of times – and the serial number.

THE STRAP

A Rolex bracelet, or strap, requires almost as much work as a Rolex movement. It consists of over 100 jointed pieces, crafted simultaneously to ensure complete uniformity. The edges are chiselled and smoothed to ensure maximum comfort. The clasp is easy to open and the length of the strap may be adjusted for a perfect fit.

A Rolex strap undergoes very rigorous tests, simulating the most stressful conditions.

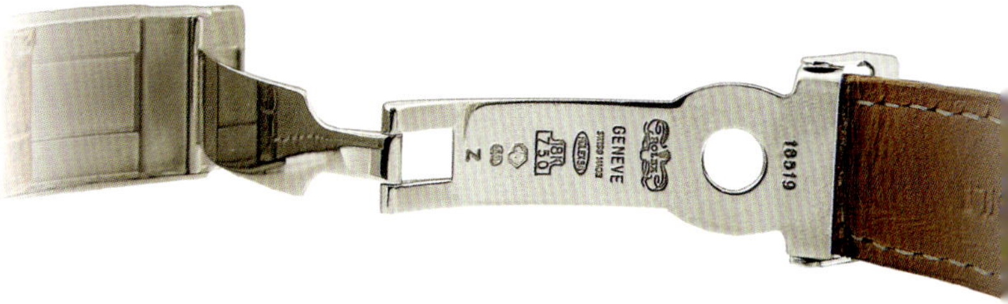

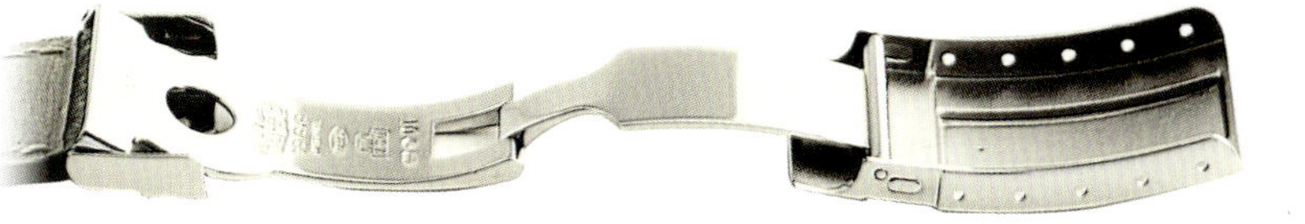

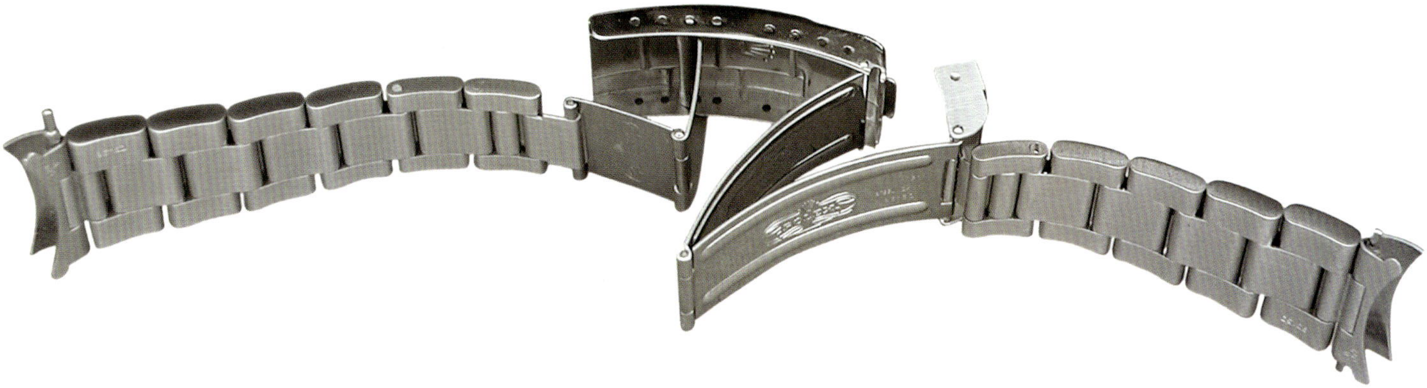

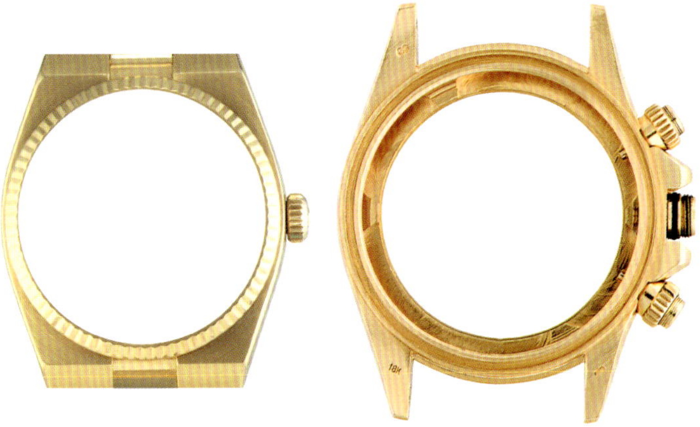

THE CASES

The design of the case is the same for all Rolex models. The winding crowns are the same for all models within the Professional and Oyster Perpetual series. The cases in the Professional series differ from those of the other series because they have protective guards on both sides of the winding crown.

THE DAYTONA PUSH PIECES

The cylindrical pushers of the cutting-edge Daytona chronograph are waterproof thanks to the screw-down system and are adorned with two parallel circles at their ends. They are made of the same material as the case and the bezel.

ROLESOR

The first watches with a bi-colour case date from the 1920s. These are mostly a combination of yellow or pink gold and white gold or platinum. The combination of gold and steel was called Rolesor, a trademark recorded by Rolex in 1933.

BEZELS

The Oyster Perpetual models differ from those of the Professional series: whereas the former feature a smooth or finished bezel, the latter – which are sports watches – present different kinds of bezels, each marked with the reference number of the model to which it belongs.

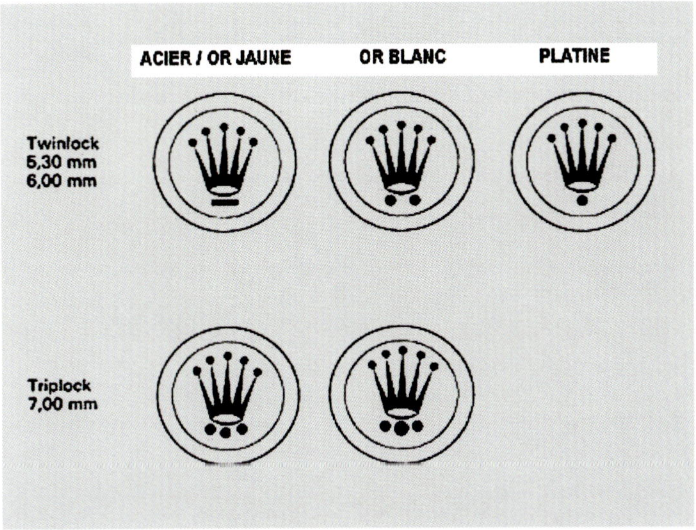

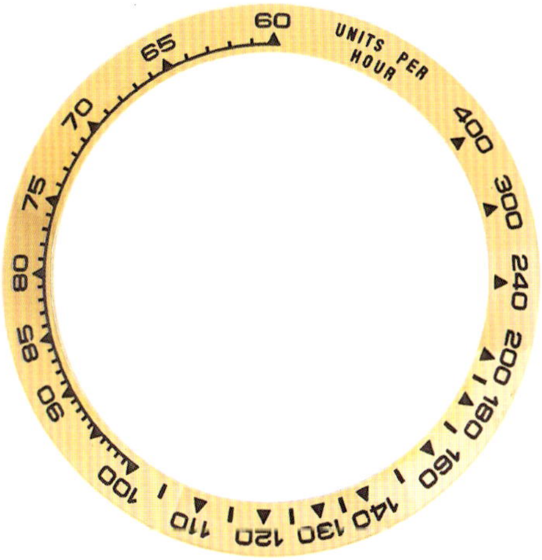

THE CROWN AND ITS SYMBOLS

A short line under the five-pointed star logo on the winding crown denotes that the model is steel or yellow gold. If there are two dots, then the watch is made of white gold, whereas a single dot indicates it of is made of platinum. In all three cases, it will be a Twinlock winding crown.

Three dots of equal size under the logo denotes the model is made of yellow gold or steel but has a Triplock crown. A larger central dot indicates a white gold watch with a Triplock crown.

UNITS PER HOUR
60 65 70 75 80 85 90 100 110 120 130 140 150 160 40
ROLEX
OYSTER PERPETUAL
SUPERLATIVE CHRONOMETER
OFFICIALLY CERTIFIED
COSMOGRAPH
DAYTONA
T SWISS
MADE T

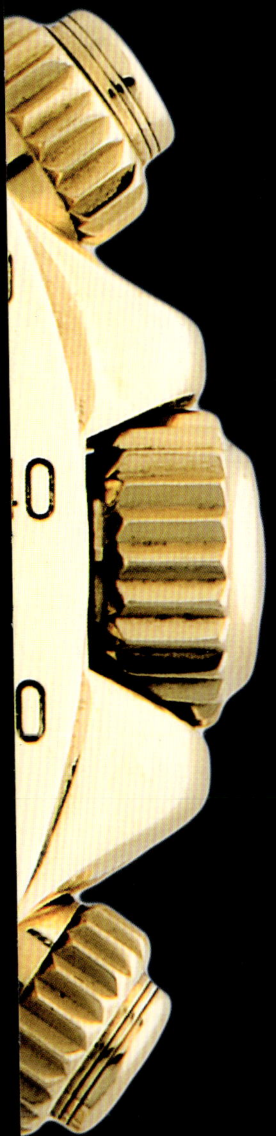

HISTORY AND LEGEND

A MODERN WATCHMAKER

March 22, 1956 was an important date for Hans Wilsdorf and the company he had founded almost fifty years earlier. This was the day of Wilsdorf's 75th birthday and at the dinner party to which all his employees had been invited, he held a speech that in many people's view was meant to clear the way for his successors. Three years after Wilsdorf's death, André Heiniger was elected CEO.

Wilsdorf's speech retraced the key events that had enabled Rolex to survive and prosper over time, and to make it through the difficult years between the two World Wars. It illustrated the company's present strategy and delivered a promise to carry on this pursuit of excellence with even greater dynamism. On this occasion, Wilsdorf emphasised what has always been the strength of all the leaders of this company: their spirit of enterprise. This spirit strives towards a modern and sophisticated form of business management, yet without ever losing sight of the distinctive priorities of watchmaking. Wilsdorf had a very clear idea as to what the essential features of a watch should be: accuracy, reliability and practicality. These are the three key points around which the Rolex strategy revolves; and through his efforts to perfect these functions, Wilsdorf became a trailblazer in the field of wristwatches.

Born in Kulmbach, Bavaria, in 1881, Wilsdorf lost both his parents at the age of twelve and was adopted by his uncles. This family situation, instead of making him insecure, strengthened his character and will. Wilsdorf completed his education with particularly excellent results in mathematics and foreign languages, and this enabled him to work abroad. He gained his first work experience as an apprentice in a company specialising in the import and export of cultivated pearls. As he himself was often to observe in the years to come, this experience afforded him considerable expertise in the field of international trade.

In 1900, Wilsdorf moved to the Swiss town of La Chaux-des-Fonds as the English-language correspondent and jack-of-all-trades of Cuno Kourten, a company specialising in the manufacture of high-quality watches.

Cuno Kourten has its own *comptoir*: the equivalent of a

Hans Wilsdorf founded Rolex in 1908. Accuracy, reliability and practicality are the golden rules that always govern the manufacturing of his watches.

modern-day showroom for wholesalers. The company had commercial ties with most areas of the world and an annual turnover of over a million Swiss francs. La Chaux-des-Fonds was a prosperous centre that vied with Geneva for market leadership in high-quality watches. The Canton of Jura, as well as the surrounding areas as far as Besançon, looked to La Chaux-des-Fonds for watches: the town's *comptoirs* drew merchants and professionals from all over the world.

This was a particularly dynamic moment for society. The turn of the century was a time of great excitement, everyday life underwent momentous changes, particularly with the appearance of the first cars, aeroplanes and bicycles. The possibility of communicating and travelling more easily fostered changes not just in people's way of living and thinking, but also in the arts, including fashion and design. Pocket watches came to be seen as a thing of the past, as the dynamic quality of modern life called for a more practical accessory, more suited to people's new needs: the wristwatch.

It was during this period that Hans Wilsdorf chose to

A pocket watch and a wallet watch from the 1920s and 1930s.

move to London and set up his own business. In 1905, together with his sister and brother-in-law, who had funded the enterprise, he established Wilsdorf & Davis, a Company for the Manufacture of Cases and the Trading of Quality Wristwatches. In his memoirs, Wilsdorf recalls that the immediate success of the company was ensured by some crucial choices: "Les Spécialités" and new products. Among the latter, the most sought-after items were travel clocks, also known as "wallet watches". These were luxury foldable leather-cased clocks that would fit in to a handbag or pocket. This type of timepiece soon became popular throughout England and beyond.

A challenge won in England

According to the registers of the Swiss customs office, watch exports were very limited during the early 20th century. Wristwatches were still regarded with suspicion at the time.

With the exception of young people, sportsmen and some female customers who saw it as a modern, avant-garde accessory, most people were rather sceptical: they thought the wristwatch was difficult to read because of its small size, as well as too fragile and too exposed to dirt, dampness and knocks to be considered reliable. Although the commercial results fell below expectations and part of the market still had to be conquered,

Two travel clocks in
the Art Deco style.

for the enterprising Wilsdorf, the wristwatch was the way of the future. The sector that unexpectedly proved to be more open to this novelty was that of fashion: the possibility of choosing a watch suited to one's clothing foreshadowed a significant development in the design of this accessory.

Confident in his choices, in the early months of 1906 the young Wilsdorf chose to travel to Bienne, near Bern, in the Jura district.

After broadening his network of contacts during his sojourn in the region, Wilsdorf issued a purchase order of many hundreds of thousands of francs to Herman Aegler, a specialist manufacturer of high-quality small watch

Operating in London from as early as 1905, Wilsdorf had access to the British colonial market.

Below and on the opposite page
A watch manufactured for the Chinese market in the 1920s, as shown by the ideogram on the dial, which translates as 'imported'. Designed as a pocket watch, it could also be set on a leather strap and worn on the wrist.

movements. This amounted to a considerable sum at the time: it was the largest order of movements Aegler had ever received. Herman Aegler had been manufacturing accurate and robust small-size movements mechanically from as early as 1902; and, despite their reduced size, he sold these movements for a reasonable price. (The two main reasons for the high cost of a quality product are its diameter and width: the smaller the watch, the more expensive it is to make.)

In London, the movements would then be mounted into cases manufactured by leading English designers – usually round cases made either of silver or 9ct, 14ct or, more rarely, 18ct gold.

The new models presented on the British market – and from there to Eastern and Far-Eastern colonies – strongly contributed to the immediate success of the enterprise. These encouraging results led Wilsdorf to make further trips to Bienne in 1906, both in order to commission new orders and to make the changes required to further increase the product's reliability.

By 1908, Wilsdorf & Davis had become the leading company in Britain for imported watches. Wilsdorf used English-made cases because the customs duty on precious materials was extremely high, which meant it was not worth importing them from Switzerland. British watches, the expression of a long-standing tradition, were products of outstanding quality that were in high demand at the time, despite their cost. Brands such as Frodsham, Dent, Benson, Smith & Son, Nicole Nielsen, Bennet and

ROLEX

ROLEX
SWISS MADE

In the 1920s and 1930s Rolex created some elegant timepieces and design watches.

Opposite page
A wristwatch with a covered rectangular case; manufactured in 1932, it could be removed from its strap and used as a pocket or purse watch.

Player manufactured extremely complex wristwatches, chronographs, tabletop clocks and travel watches. These typically sturdy items stood out for their movements as well as for the high quality of their dials and cases.
Swiss products proved competitive because of their quality and affordability, which enabled the Wilsdorf & Davis watches to conquer the British as well as international markets within a few years.
Another event that contributed to the success of the company's wristwatches was the extendable bracelet. Practical and resistant, this highly versatile element – which fitted all sizes – helped turn the watch into another standard accessory for men, alongside cufflinks, tie clips and lapel pins.
The cases varied in terms of shape and design. In many ways they were highly innovative: not just round, but also oval, square, octagonal or rectangular; some also featured a covered dial (*savonnette*) – one variant being the "bull's eye" dial (with an opening at the centre of the cover, which made it possible to read the watch while protecting the dial) – with a screw-down bezel, often in combination with a screw-down back, to hermetically seal the part of the case enclosing the movement. This model foreshadowed the Oyster watch.
One of the factors that prevented Wilsdorf & Davis from establishing itself as a brand was the typically British practice of retailers branding dials with their own logo rather than that of the watchmaker. This, along with Wilsdorf's wish to prove himself directly on the market, is what led him to register the Rolex brand in the Swiss Federal Trademark Office at 8am on 2 July 1908.

A "royal" brand

A number of theories have been put forward to explain the choice of the name Rolex, but none was ever confirmed by Wilsdorf himself. The most plausible hypothesis – and

The first Oyster watch with a rounded square case from 1926 (left) and the first Oyster Perpetual from 1931 (right).

the one which best fits with Wilsdorf's personality – is that which invokes his role as a brilliant pioneer and successful marketing strategist. Wilsdorf simply wanted a name that would be easy to pronounce in all languages: a simple, brief and incisive one. What better name could be found than "Rolex"? From 1925 onwards, the logo also came to feature a royal crown above the name, and this soon became the symbol of these products of outstanding quality. The brand was registered with trademark no. 24001. The company's headquarters were given as La Chaux-des-Fonds. The company continued to operate from there until 1912, when it moved to Bienne, and then to its final location in Geneva in 1919, with the business name of "Manufacture des Montres Rolex".

The watches were manufactured and tested in Bienne and then re-tested and prepared for despatch in Geneva. Wilsdorf did not wish to abandon the British market, so he struck a compromise with local retailers in order to bypass the prohibition to mark watch models with the manufacturer's name, and hence finally make the Rolex brand known. Wilsdorf concluded an agreement whereby for each box of six watches delivered to Britain one would bear his mark on the dial. Though this started as just one watch, it later became two, then three, and so on. Starting in 1925, with the introduction of the crown logo, as the result of an intense advertising campaign, the Rolex name also finally became well-established in Britain, so that there was no need to come up with any further agreements. It took Wilsdorf no less than 17 years to establish his brand, and this was only possible thanks to his perseverance, extensive knowledge of the market, and brilliant marketing ideas.

Innovation through research and design

At the turn of the 20th century, the design of wristwatches was strongly influenced by that of pocket watches: they had a round shape, with wire lugs for the strap. The case would have a large winding crown at 12 o'clock or 3 o'clock and the body would usually be in chrome or nickel-plated or burnished iron – only rarely of more precious metals. Often a metal mesh would be added to protect the dial and glass from any damage.

ROLEX
OYSTER
ULTRA PRIMA
SWISS MADE
ROLEX OYSTER
PERPETUAL
SELF WINDING
SWISS MADE

The Prince model first produced in the late 1920s. The geometrical and linear design of the dial reflects Art Deco aesthetics.

The young Wilsdorf too began by manufacturing round watches with a winding crown at 3 o'clock. From the end of the first decade of the century, however, the experimental research of the Belle Époque inspired new shapes, adding a more distinctive, classy touch. The wristwatch became a status symbol.

With its aesthetic principles and geometrical lines, Art Deco influenced all fields, including that of watchmaking. Square, hexagonal, rectangular or pointed cases were produced, which stood out for their variety and quality. This early experimental research later led to the introduction of the famous *tonneau* shape, which is still used for many Rolex models, in both the men's and the women's collections.

The dial was typically of white enamel with large Arabic – or more rarely Roman – numerals, coated with phosphorus for luminescence. Phosphorus would also be applied to the hands. Black enamel dials with white or silver metal hands were also manufactured. Almost all models had an auxiliary dial for the seconds.

These years witnessed the launching of the Prince model

A brilliant marketing man, Wilsdorf launched a number of high-profile events, such as Mercedes Gleitze's swim across the Channel to test the waterproof qualities of the Oyster model. A forward-thinking entrepreneur, Wilsdorf also relied on quality advertisements to promote his company.

and its counterpart Princesse, available in various sizes and with various combinations of different metals. This model was to become the first Rolex icon.

Clearly, technological progress also contributed to the development of the company.

In 1926, Rolex launched the first Oyster case, with a screw-down bezel, cover and winding crown. In a canny move, a year after patenting the Oyster watch with a water-tight case, Wilsdorf sponsored an event that had global resonance. On 21 October 1927, Mercedes Gleitze swam

ROLEX OYSTER
SWISS MADE

Some watches from the 1920s feature a round dial set within a geometric case.

Opposite page
A model with a hexagonal case.

A Rolex Datejust with reference 6105, introduced in 1952.

across the chilly waters of the English Channel wearing an octagonal, 31x27mm, 9ct-gold Oyster Precision watch on her wrist. On her arrival, the watch was checked and declared to be "working perfectly".

Another big innovation came in 1931: the problem of re-charging was solved through the Perpetual device, which enabled the watch to self-wind through the movement of its wearer's arm and thus remain constantly charged. In the 1940s and 1950s other models were developed that were soon destined to become classics, such as the Datejust, Submariner, and GMT-Master.

In 1960 yet another event occurred that caused quite a stir, and not just in the watchmaking sector either. The Deep Sea Special model, especially designed to endure high pressure underwater, was strapped to the outside of the self-propelled deep-sea submersible *Trieste*, which dived in the Mariana Trench, off the island of Guam, to a depth of 10,916 metres. To enable the watch to endure a pressure of over one ton per square centimetre, an extremely resistant steel case was designed, with a convex glass a few centimetres thick. When the watch was brought back to the surface, Jacques Piccard, the head of the expedition, confirmed that it was "working perfectly".

The story continues

Hans Wilsdorf passed away in 1960. Three years later, André J. Heiniger was elected president and CEO of Rolex, and contributed to Rolex's lasting international fame. Previously in charge of the Argentinian market and Hans Wilsdorf's favourite employee, Heiniger had attained this position at the age of 42. The management change was a positive event for Rolex. A man with considerable commercial expertise, Heiniger strengthened Rolex's image not just by continuing to manufacture products of the highest quality, but also

In 1960, a Deep Sea Special was strapped to the side of the bathyscaphe Trieste, which plunged into the Mariana Trench. The watch kept running even at a depth of 10,000 metres.

Opposite page
Rolex ads from the 1960s.

by expanding the administrative sector of the business management.

Heiniger further renovated the distribution system through a more accurate selection of agents and a stricter evaluation of competences, sale prices and post-sale services, developing a more global marketing approach for the company.

One of the most evident and striking achievements of Heiniger's management was that it further increased the already huge credibility that the Rolex brand enjoyed in both the cultural and commercial sphere – even during the daunting 1970s.

The development of quartz watches and the invasion of accurate and cheap Japanese products posed a number of challenges to the Swiss watch industry as a whole. Nevertheless, thanks to the quality of its products and its rigorous customer service, Rolex managed to preserve the trust of foreign markets, attaining the enviable status of a financially stable and reliable company.

During Heiniger's management, Rolex launched a number of new products and improved existing ones, setting itself on course to become the world's leading watchmaker in the 1980s.

Among the most striking models created during those years were the famous Daytona Chronograph (launched

ROLEX
COSMOGRAPH
DAYTONA

- Stopwatch timing to 1/5 second
- Minute and hour recorders
- Extra rugged, patented case, made in Geneva
- Amagnetic
- 17 jewel super shockresisting movement
- Unbreakable mainspring
- Sweep second hand
- Crackleproof crystal
- Available in white/black or black/white dial combinations

ROLEX - OFFICIAL TIMEPIECE PAN AMERICAN AIRWAYS

ROLEX - OFFICIAL TIMEPIECE DAYTONA INTERNATIONAL SPEEDWAY

THE ROLEX WATCH COMPANY LIMITED, (FOUNDER H. WILSDORF), GENEVA, SWITZERLAND

ADEN, AUCKLAND, BOMBAY, BUENOS AIRES, COLOGNE, DUBLIN, HAVANA, HONGKONG, JOHANNESBURG, LONDON, MELBOURNE, MEXICO CITY, MILAN, NEW YORK, OSAKA, PARIS, SÃO PAULO, SINGAPORE, SYDNEY, TOKYO, TORONTO.

PRINTED IN SWITZERLAND

get up and go

ROLEX
COSMOGRAPHE

- Cronografo preciso ad un quinto di secondo
- Registratore di minuti e di ore
- Cassa robusta e brevettata fabbricata a Ginevra
- Antimagnetico
- Movimento 17 rubini, scappamento anti-urto
- Molla infrangibile
- Sfera dei secondi al centro
- Vetro che non si fende
- Quadrante bianco con registratori neri o quadrante nero con registratori bianchi, a scelta.

Rif. 6239

ROLEX — CRONOMETRO UFFICIALE DELLA «PAN AMERICAN WORLD AIRWAYS»

MONTRES ROLEX S.A. (FUNDATORE H. WILSDORF) GINEVRA, SVIZZERA

ADEN, AUCKLAND, BANDUNG, BANGKOK, BOMBAY, BRUSSELLE, BUENOS AIRES, CARACAS, COLONIA, DUBLINO, HONG KONG, JOHANNESBURG, L'AVANA, LONDRA, MADRID, MANILA, MELBOURNE, MESSICO, MILANO, NUOVA YORK, OSAKA, PARIGI, SAO PAULO, SINGAPORE, SYDNEY, TOKIO, TORONTO.

STAMPATO IN SVIZZERA

COSMOGRAPHE

ROLEX

In 1953 Rolex created the first wristwatch for deep-sea diving: Submariner, an automatic watch designed to be water resistant up to 100 metres. In 1992, the Yacht-Master was introduced, a sports watch much appreciated by yachtsmen.

in 1963), the "610" Sea-Dweller (1967), the Submariner (1969), the Oysterquartz (1974), the "4000" Sea-Dweller (1978), the automatic Daytona Chronograph (1988), and the Oyster Perpetual Yacht-Master (1992) – a woman's version of which was developed in 1994.

In 1976, André Heiniger established the Rolex Awards for Enterprise with the aim of promoting the entrepreneurial spirit around the world. This award is bestowed for the most interesting contributions made in the field of applied science, inventions, explorations, discoveries and ecology. Through this acknowledgement, Heiniger and Montres Rolex SA have provided concrete support to hundreds of worthy men and women. Through Heiniger and the Wilsdorf Foundation, Rolex also provides help and assistance to the poor and the blind, as well as scholarships for students at Geneva University.

André Heiniger left Rolex in 1997. The board of directors appointed his son Patrick Heiniger as CEO in 1992. Patrick – who earned a Law degree from Geneva University in 1977 – followed Rolex's business closely and travelled extensively in order to gain detailed knowledge

Always abreast of the latest communication strategies, Rolex created new advertisements for each launch.

of the way the company operated around the world. He had first joined Rolex as Marketing Director in 1986. He grounded his professional training on the spirit of enterprise and acquired extensive commercial experience, along with an ability to relate to people, which he put to the service of Rolex. With discretion and enthusiasm - virtues typical of the Heiniger family - Patrick Heiniger remained true to the company's philosophy. He fostered new dynamism in both the technological and human sector. He oversaw the development of new structures in Geneva and the inauguration of the new Rolex building in 1995. In 2002, he established the Rolex Mentor and Protégé Arts Initiative, a programme that each year brings together internationally acclaimed artists active in different fields and young talents, for them to work together on artistic projects or create artworks. Under Patrick's management, Rolex purchased a number of supplier companies and centralised the different sectors of the group by concentrating them in the new building, which was inaugurated in Plan-les-Ouates, near Geneva, in 2005. Patrick Heiniger stepped down at the age of 58 on 16 December 2008.

Bruno Meier, who was appointed CEO late in 2008, is the former Chief Financial Officer of Rolex. Prior to that, he had served as the director of Deutsche Bank Switzerland, as global COO of Deutsche Bank Private Banking, and as head of the Risk Management Unit of ZKB. Meier stepped down after only a short time for personal reasons. Gian Riccardo Marini was appointed CEO of Rolex SA on 4 May 2011, with the aim of further promoting the

industrial and commercial growth of the company at a global level. Marini is the former Managing Director of Rolex Italia S.p.A. Then in April 2014 Jean-Frédéric Dufour became the head of Rolex SA.

And so the story continues...

Rolex today

Rolex belongs to the Hans Wilsdorf Foundation, a Swiss private trust established by Hans Wilsdorf himself in 1945. Having lost his wife in 1944 and having no direct heirs, Wilsdorf chose to donate all his shares to the foundation, which has been managing the company's revenue since then.

According to the statute of the trust, part of this revenue must be reinvested in technological research, to constantly improve the products. Another substantial share is set aside for social work or the funding of artistic projects. The Rolex headquarters are in Geneva, where the administrative and marketing offices are also located. The rigorous functioning tests, the monitoring

The time-honoured Rolex headquarters in Geneva.

of certain models and the pre-sale services are all run in Geneva, while the main manufacturing plant is in Bienne. According to market analysts and the Swiss investment bank Vontobel, Rolex is the best-selling watch company in the world, with sales estimated at 4,500 million Swiss francs in 2012, equal to 3,700 million euros. This is only a theoretical estimate, however, since the company's policy is not to disclose financial data.

These production volumes also make Rolex the biggest gold-processing company in Switzerland.

In accordance with the statute of the Hans Wilsdorf Foundation, the company is not listed on the stock market. In agreement with the moral legacy of its founder, it operates in the pursuit not merely of profit, but of the overall growth of the value of its brand.

On celebrity wrists

Aside from having founded the greatest *haute horologerie maison* in the world, Hans Wilsdorf and his successors proved capable of making their products a legend. Over time, the *maison* has successfully associated its watches with remarkable people and the events connected to them. A Rolex watch accompanied Mercedes Gleitze on her swim across the Channel, Sir Edmund Hillary on his ascent of Mt Everest, Jacques Picard on his deep-sea plunge, and Sir Malcolm Campbell on the nine-times he set new speed records on board his *Bluebird* between 1924 and 1935. The list of celebrities wearing Rolex watches includes actors and musicians such as Paul Newman, Sean Connery, Eric Clapton and Placido Domingo, politicians such as Fidel Castro, Winston Churchill and Dwight D. Eisenhower, and famous sportsmen like Michael Jordan, Cristiano Ronaldo and Roger Federer – exemplars of constant commitment, as well as great skill.

Rolex has always promoted pioneering ideas and innovative thinking, helping people achieve great feats. The value of dreams and importance of creativeness are concepts so central to the *maison* that they have inspired the philosophy behind the Rolex Mentors and Protégés Arts Initiative. Launched over a decade ago, this programme seeks to preserve the traditional relation between master and apprentice, so as to ensure the transference of

Since 2013 Rolex has been a Formula 1 sponsor.

the world's artistic heritage from one generation to the next. Involved in the initiative are leading representatives of seven disciplines within the arts: architecture, dance, cinema, literature, music, theatre and the visual arts; each is 'entrusted' with young budding talents, often from different continents and cultures. These pairs of artists – such as Lin Hwai-min and Eduardo Fukushima for dance and Margaret Atwood and Naomi Alderman for literature – are given the opportunity of working together as they please for one year of creative mutual collaboration.

Aside from funding philanthropic projects, the Geneva-based *maison* also sponsors the world of culture and sport. A long-time supporter of sailing, tennis and golf events, in 2013 it became a main Formula 1 sponsor – a tribute to its well-established link with the world of car racing. The company has also launched the Rolex Grand Slam of Show Jumping, a new competition in the equestrian world.

Timeless luxury

Perfectly recognisable, since their design has not changed over time, Rolexes are more than just watches: they are coveted items for many people, and an object of veneration for enthusiasts and collectors.

The value of these iconic watches does not merely endure over time but grows, as is also shown by the outstanding sales results accrued over the years in the most important auction houses around the world: Antiquorum, Christie's and Sotheby's.

One important goal was reached in 2011 when at a Christie's auction a split-second chronograph from 1942 – reference 4113 – fetched over one million Swiss francs (CHF1,035,000).

In May 2013 another split-second chronograph, manufactured in 1942, was auctioned off for 1,107,750 Swiss francs, beating the record previously set by the same reference. Over the course of the same event, another record was set for the Rolex Submariner, a model that was sold for 519,750 Swiss francs.

Again in 2013, this time at Antiquorum's, a gold Daytona Paul Newman (ref. 6263) with a rare lemon-yellow dial and contrasting black patterns was purchased for the sum of 841,300 Swiss francs. The year

2013 proved an exceptional one, as two other auctions ended with thrilling results. In the November, to celebrate the 50th anniversary of the iconic Daytona model, Christie's organised the "Lesson One" event in Geneva, auctioning off some of the most sought-after watches by collectors around the world. The event centred around 50 Rolex Daytonas, some of which – formerly owned by celebrities who had chosen them as their favourite watches – had never been put to auction before.

On this occasion, a Paul Newman Rolex Daytona from 1969 – reference 6263 – was assigned to a collector for 989,000 Swiss francs. A customised model manufactured for the Sultan of Oman Qaboos Bin Said Al Said in 1974 was then auctioned off at the price of 785,000 Swiss francs.

It was late in 2013, however, when Rolexes gave their best. At Christie's auction on 17 December, a rare model from 1950 (ref. 8171) – nicknamed "Padellone" (Big Pan) on account of its oversized case (38mm) – was sold for $1,145,000, four times the starting bid.

It is certainly difficult to explain the prices fetched by these watches simply on the basis of their fine craftsmanship. Some of these pieces belonged to celebrities, while others are extremely rare. What increases their value, however, is the legend, desire and idea of exceptionality that Rolex has always been very good at nourishing.

12
ROLEX
9
3
SWISS MADE

ICONS

The steel Prisoner Watch (ref. 3525) was also marketed among British officers, who could order them in German prison camps during the Second World War.

Produced in limited editions or for celebrities, some timepieces have left a permanent mark on the history of Rolex.
An iconic Rolex came into being in 1928: the Prince model, a rectangular watch with a double dial. It was also known as the "Doctor Watch" because of the neat separation between the seconds dial and that of the hours and minutes, which made it easier for physicians to take a patient's pulse.
Other famous models include the first chronographs, which were marketed between the 1930s and 1950s and are exclusively technical, functional objects: they were designed to be worn by engineers, chemists, architects, officers and sportsmen as a work tool that would ensure a high degree of accuracy and reliability. For this reason, Rolex chronographs were only manufactured in very small numbers – which is what makes them all the more valuable today. The watches produced for the Italian

Split-second anti-magnetic steel chronograph; just twelve copies were manufactured in 1942. This model is highly sought after by collectors. Ref. 4113.

Navy in 1938 instead stem from the collaboration between Rolex and the Florence-based G. Panerai & Figlio company. Panerai designed the watch, while Rolex manufactured the crown, case and movements.

Particularly rare and sought after by collectors throughout the world are the twelve split-second chronographs (with two independent sub-dials for the seconds) manufactured in 1942.

Another limited production is that of the Prisoner Watches, which British officers captured by Germans could purchase during the Second World War. These watches would be brought to the prison camps by the Red Cross and given to the officers for free, with the provision that they would pay for them in instalments at the end of the war.

In 1947 Rolex introduced a model with a triple date and chronograph. It takes its name from the complete calendar (dato) and three subsidiary chronograph dials, known as "compax".

A Radiomir Panerai model created by Rolex for the Florence-based G. Panerai & Figlio company in 1938. This extremely valuable model fetched $82,600 at an auction in 2007. Ref. 3646.

Right
The chronograph that collectors commonly refer to as the "Pre-Daytona". Manufactured in both steel and 14ct or 18ct-gold versions, it is listed in Italian Rolex catalogues from 1964-1967. Ref. 6032.

In 1949 Rolex created some models with the moon-phase function. Very few of these watches were ever manufactured. Ref. 8171, in pink gold, has a leather strap.

Right
A ref. 6062 from 1950 with an Oyster case, a pink-gold bracelet and a restored dial.

Chronograph from the 1950s, ref. 6236, with pink-gold case and bracelet; referred to by collectors as the "Jean-Claude Killy".

Opposite page
What distinguishes the Prince model is not just its rectangular and convex shape, but the fact that it has a separate seconds dial.

The year 1951 witnessed the appearance of the Dato Compax, which collectors refer to as the "Jean-Claude Killy" model – after the famous skiing champion of the 1960s, who was a great Rolex enthusiast and was particularly fond of this model.

In 1950 Rolex, which up until then had produced few complicated watches, opting for simple but resistant designs instead, started developing some intricate moon-phase mechanisms. Between 1950 and 1953, 350 yellow-gold moon-phase watches were manufactured, as well as 50 pink-gold watches and a very limited number in steel. On account of their rarity, these pieces are highly sought after by collectors the world over.

ROLEX
PRINCE
SWISS MADE

TIMELINE

1908

Hans Wilsdorf establishes the Rolex Watch Co.

1910

A Rolex watch is certified as a chronograph.

1914

Rolex is awarded the first class "A" certificate from the Kew observatory.

1925

The distinctive five-pointed crown is registered as the official Rolex logo.

1926

Rolex patents its first screw-down winding crown and its first waterproof Oyster case.

1927

The English swimmer Mercedes Gleitze swims across the English Channel wearing a Rolex Oyster. On her arrival, the watch is found to be running perfectly.

1928

Rolex launches the first Prince models, nicknamed "Doctor Watches".

1930

Rolex launches the first chronographs with a push-back case, with a single push piece (ref. 2022-2023-2303) or two set at the sides of the winding crown (ref. 2508).

1931

The first Rolex Perpetual automatic mechanism is patented. This automatic movement is mostly associated with a model that has made history: an Oyster commonly known in Italy as "ovetto" ("little egg").

1938

The first watches commissioned by the Italian Navy are ready for despatch. Designed by the Florence-based G. Panerai & Figlio company, they feature crowns, cases and movements signed by Rolex.

1942

Rolex creates ref. 4113: a split-second chronograph, of which only twelve copies were ever produced.

1945

The Rolex Datejust is presented at the Basel Fair: the first automatic watch with a calendar aperture and automatic date changing.

1947

Rolex launches references 4767 (with an Oyster case) and 4768 (with a push-back case): two new chronographs with totalizer function and date calendar better, known as "Dato-Compax".

1949

Rolex introduces ref. 8171, with a moon-phase complication.

1950

Rolex creates ref. 6062, the zenith of its most important collections. With moon-phases and a date calendar, between 1950 and 1953 Rolex produced 350 yellow-gold examples, 50 pink-gold, and a limited number of steel copies.

1953

Sir Edmund Hillary and Tenzing Norgay conquer the peak of Mt Everest; Hillary is wearing a Rolex Oyster Perpetual model, the Explorer. The maison's first professional models are also presented: the Turn-O-Graph (ref. 6202) and the Submariner (ref. 6204).

1954

Birth of the Rolex GMT-Master (ref. 6542), displaying two time zones simultaneously, and of the anti-magnetic Rolex Milgauss model. The Cyclops magnifying lens is introduced for models with a date calendar.

1956

The Rolex Day-Date makes its appearance at the Basel Fair; at the same time, the new President bracelet makes its début. The Milgauss is introduced.

1957

Birth of ref. 5500, destined to remain on the market until 1990. This reference was introduced under various names – "Air King", "Metropolitan", "Explorer" and "Everest" – before establishing itself under the name of "Air King".

1959

The Rolex Submariner (ref. 5512) introduces the use of guards to protect the winding crown.

1960

On 23 January the Deep Sea Special resurfaces without any signs of damage after having been strapped to the side of the submersible Trieste and plunged to a depth of 10,916 metres.

1962

Rolex presents a new chronograph in steel, 14ct-gold and 18ct-gold versions: ref. 6238, known as the "Pre-Daytona".

1963

Birth of the first Rolex Cosmograph Daytona chronograph (ref. 6239).

1965

Rolex introduces screw-down chronograph pushers with the Cosmograph Daytona model (ref. 6240).

1969

Birth of the first Submariner with a date calendar, Submariner Date (ref. 1680).

1971

Birth of Explorer II, equipped with a fixed bezel subdivided into the 24 hours and with a subsidiary hour hand.

1977

Rolex presents the Oysterquartz series, the first watch with a quartz movement manufactured by the company.

1978

Rolex launches the Sea-Dweller 4000 (ref. 16660), waterproof to a depth of 1,220 metres (4,000 feet) – exactly double the depth of the previous series (ref. 1665).

1979

Rolex launches the GMT-Master, a steel model with the quick date change function (ref. 16750); this is immediately followed by a steel/gold model (ref. 16753) and an 18k-gold model (ref. 16758).

1983

Issuing of the new Explorer II (ref. 16550), available with a black, white or cream dial. Rolex creates the Gmt-Master II, which can display three time zones (ref. 16760, steel).

1988

Birth of the first self-winding Rolex chronograph (ref. 16520) with a Zenith movement.

1990

Rolex develops the new Calibre 3000, self-winding and with seconds at the centre.

1992

Rolex launches the Yacht-Master.

2000

Rolex presents the new Cosmograph Daytona models with a chronograph movement fully manufactured by the company.

2003

The Submariner celebrates its first 50 years and for the occasion Rolex introduces the new steel model with green ring and "maxi" hour-markers in white gold.

2006

Launching of the GMT-Master II with a green or black dial.

2007

Launching of the new Yacht-Master II, Milgauss and Air King models.

2008

At the Basel Fair, Rolex presents the Sea-Dweller Deepsea model – waterproof up to 3,900 metres – along with new Oyster Perpetual 36mm and Submariner Date models, and the first Everose gold version of the Daytona.

2010

Launching of the 904L steel Submariner with black Cerachrom ring (ref. 116610 LN).

2011

At the Basel Fair, Rolex presents the new Oyster models: Oyster Perpetual Explorer II, Oyster Perpetual Yacht-Master II, Oyster Perpetual Cosmograph Daytona, Oyster Perpetual Datejust Special Edition, and Oyster Perpetual Lady Datejust.

2012

Presentation of a new Oyster model: the Oyster Perpetual Sky-Dweller.

2014

Presentation of the 904L steel Oyster Perpetual GMT-Master II and of the platinum Oyster Perpetual Cosmograph Daytona.

ROLEX
COSMOGRAPH
DAYTONA

THE MODELS

OYSTER PERPETUAL | 1931

The appearance of the Oyster Perpetual marked a real revolution in the history of watchmaking.

The model was given this name because it featured the first waterproof case, which was developed thanks to the invention of the screw-down crown in the 1920s. In 1926 the company patented this technology and started marketing the watch. The switch from the Oyster to the Oyster Perpetual was made in 1931, the year in which Rolex patented the "Perpetual" self-winding rotor system. This made it possible to charge the watch through the movement of the wearer's arm, without having to manually charge it every day. The Perpetual mechanism consisted of a rotor that through clockwise rotation ensured accurate and reliable charging.
Today the Oyster Perpetual is part of the Rolex's Special Collections and is equipped with an Officially Certified Superlative Chronometer.

The Rigid Hooded version of the Oyster Perpetual features covered lugs, which lend the watch a more solid appearance. Ref. 3599.

Opposite page
The Oyster Perpetual comes with different cases and dials: ref. 6100 has a smooth bezel, ref. 6285 a graduated one; the blue-dial version of the watch is very rare.

SWISS
ROLEX
OYSTER
OFFICIALLY CERTIFIED
CHRONOMETER
SWISS MADE

Particularly rare and prized are Oyster Perpetual watches with a uniform or cloisonné enamel dial, dating from the 1950s. Ref. 6100.

ROLEX
OYSTER PERPETUAL
25 YEARS SERVICE
Coca-Cola
SUPERLATIVE CHRONOMETER
T SWISS T

ROLEX
OYSTER PERPETUAL
SUPERLATIVE CHRONOMETER
OFFICIALLY CERTIFIED
T SWISS MADE T

In 1962, new references with bezels marked by different finishes were added to the Oyster Perpetual collection. Ref. 1002 has a smooth bezel.

Opposite page
Ref. 1003 has a guilloché bezel. The Oyster Perpetual (ref. 1003) marked with the Coca Cola logo was auctioned for 12,980 Swiss francs in 2007.

DATEJUST | 1945

In 1945, to mark the 40th anniversary of the founding of Wilsdorf & Davis, the company from which Rolex sprung, the Datejust model was launched.

A distinguishing feature of this model is its automatic movement with a rotor that is visible through an opening on the dial at 3 o'clock.

The first model was manufactured in yellow gold, except for a very small number in pink gold. Three years after its launch, the Rolesor version was introduced in steel and gold, and later just steel. The name Datejust, however, only started appearing on the dial in the early 1950s.

In 1954, in Basel, the *maison* presented the Cyclops lens, which was set over the calendar to enlarge the date. While on the first models, the calendar would start moving a few hours before midnight, in 1955 a new mechanism made the change of date instant.

The Datejust models boast a COSC certification attesting to their accuracy as chronographs.

Featured here is the classic steel version of the Datejust II model, made even more reliable by the addition of the Parachrom spring and anti-shock Paraflex device (ref. 116300).

Opposite page
This model is also available in a pink-gold jewel version (ref. 116285) with a striped black-enamel dial studded with 262 diamonds. The bezel is set with 60 baguette-cut diamonds and 120 brilliant-cut ones.

ROLEX
DATEJUST
28

A series of Rolex Datejusts. Featured here are references 6075 and 1601 in yellow gold with a knurled bezel – the former from 1959, the latter from 1950.

Opposite page
Reference 6304 with a steel case and smooth bezel, from 1955.

ROLEX
OYSTER-PERPETUAL
DATEJUST
30
OFFICIALLY CERTIFIED
CHRONOMETER
SWISS

EXPLORER | 1953

On 29 May 1953 New Zealand mountaineer and explorer Sir Edmund Hillary, together with the Nepalese Sherpa Tenzing Norgay, reached the highest peak in the world: Mount Everest.

Both were wearing a Rolex Oyster Perpetual - to be more exact, a prototype of the Explorer specifically designed for the expedition and featuring a leather strap instead of the steel bracelet that was used for the later version launched on the market. Rolex saw the ascent of Mt Everest as the perfect occasion on which to test its watches under extreme conditions (sharp changes of temperature and pressure) and as an important promotional launch. In the same year, the company registered the Explorer trademark and started to market the model.

In 1967, a black dial with white numerals was introduced to improve readability, and this remained in use for all later versions, except references 16550 and 16570. The accuracy of all these watches is guaranteed by the COSC certification.

In 1971, Rolex introduced the Explorer II model, capable of ensuring even greater accuracy, lightness and endurance.

Today, the Explorer II is made from 904L steel, a highly resistant and particularly shiny alloy. The dial is easy to read both in the daylight and at night thanks to its eight-hour luminescence, the size of the hands, and the orange 24-hour hand used for the second time zone. An Oyster case makes the watch waterproof.

A spin-off from the Oyster Perpetual, Explorers differ from their precursors with respect to the dial. Early Explorers have a honeycomb dial (ref. 6298), also in a cream colour (ref. 6150).

ROLEX
OYSTER PERPETUAL
EXPLORER
ROLEX
OYSTER PERPETUAL
PRECISION
EXPLORER
ROTOR
SELF-WINDING
SWISS

In 1971, Rolex renewed the Explorer range by introducing the Explorer II models, with an even easier to read dial. Ref. 1655 is set with a large orange hand indicating the 24 hours; this continues to be a feature of present-day models.

Opposite page
Ref. 16550 is offered in three different dial colours: white, black or cream. The cream version is rarer and much sought after by collectors.

24
2
4
6
8
10
12
14
16
18
20
22
ROLEX
OYSTER PERPETUAL DATE
PLORER II
16
SUPERLATIVE CHRONOMETER
OFFICIALLY CERTIFIED
T SWISS < 25 T

SUBMARINER | 1953

The Submariner confirmed the professional competences of Rolex technicians by allowing divers to measure their immersion time.

The first model, presented at the Basel Fair of 1954, is waterproof to a depth of 100 metres. The distinguishing features of the Submariner are the Oyster case, which makes it waterproof, and the rotatable bezel, designed in such a way as to only move in an anti-clockwise direction, to prevent divers from making any mistakes when recording their time.
In the 1970s, the Twinlock winding crown was replaced by the Triplock crown, to make the watches even more waterproof. All these elements have contributed to improving the performances of the Submariner, which is now pressure-resistant up to a depth of 300 metres.
The Date version of the Submariner, introduced in 1969, presents a blue, black or green bezel made of Cerachrome, a material highly resistant to corrosion. The strap on the Date models is fitted with a Glidelock clasp through which it can be adjusted and extended by up to 18 millimetres, enabling divers to wear the watch over their wetsuits.

Ref. 1680, launched in 1969, is marked by a gold case and a blue dial and bezel. It is the first gold Submariner with a date calendar. The gold bracelet features a fliplock clasp that makes it adjustable in length.

Early versions of the Submariner (featured here is Ref. 5508) had no protection guards on the two sides of the winding crown. Ref. 5508 is waterproof to 100m/330ft.

References 6204 and 6205 of the Submariner came to be known as the "James Bond" models among collectors, as in the films Sean Connery wears a Submariner without a date calendar and with a strap in a grey and white striped military fabric. Featured here is Ref. 6204.

Opposite page
Ref. 5510 (left), first introduced in 1958, is waterproof to 200 m. It is nicknamed "Coroncione" because of its 8 mm crown.

Ref. 5517 (right) was launched in the 1970s. The watch presented here was manufactured for the Royal Navy.

ROLEX
OYSTER PERPETUAL
SUBMARINER
ROLEX
660ft = 200m
SUBMARINER

1998 Ref. 16613, in steel and yellow gold with a gold dial.

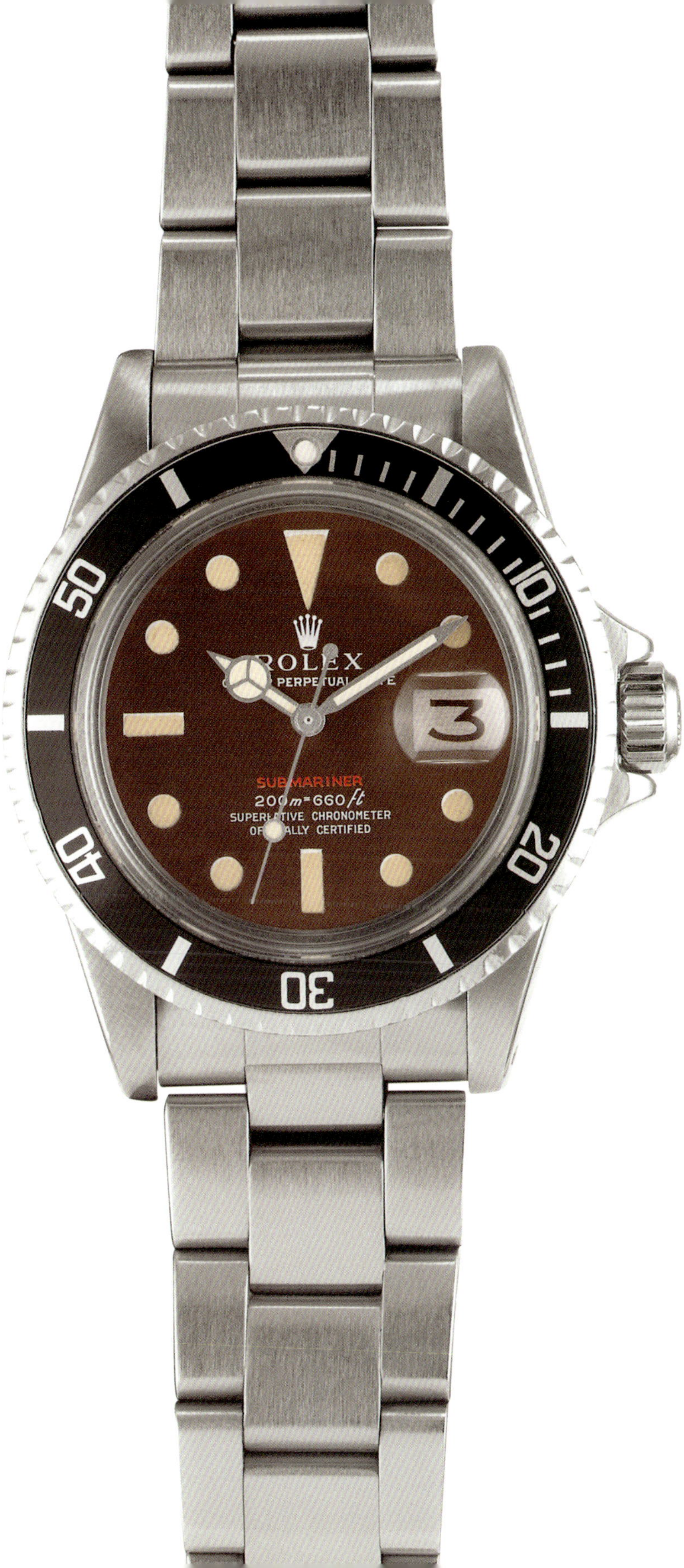

One unusual model is the brown-dial watch with an outer minute scale (ref. 1680), from 1974.
In 2003, Rolex celebrated the 50th anniversary of the Submariner by launching a version with a green outer ring (ref. 16610 LV).

GMT-MASTER | 1954

The GMT-Master sprang from the collaboration between Rolex and Pan-Am Airlines, the first airline to provide intercontinental flights without a stopover.

To help pilots and flight attendants avoid any confusion related to the change of time zones, Pan-Am shared with the Geneva-based company its concern to be able to know both the local time in the place of landing and that at the point of departure. In response to this, Rolex launched the GMT-Master model in 1954. It displays the second time zone through a fourth hand complication that indicates the time on a graduated bezel marked 0-24. The rotating bezel has a luminous coating and is coloured red for the daytime hours, and blue for the night-time. At the dawn of the 1970s, a black bezel was introduced as an alternative to the red/blue one. In 1983 Rolex created the GMT-Master II, which displays three time zones. This model is equipped with a 24-hour hand and a 12-hour hand, which can be set separately in such a way as to mark a second time zone. When the time indicated by the hour hand is shown on the dial, the time indicated by the 24-hour hand can only be read by looking at the bezel. In order to accurately indicate a third time zone, the bezel must rotate without making any intermediate stops, with a limited number of 24 possible positions. This is ensured by a spring enabling individual symmetrical movements in both directions.

Ref. 1675, manufactured in pure 18ct gold, was launched on the market in 1960. In the early 1970s a Rolesor (steel and gold) version appeared.

ROLEX
GMT-MASTER

ROLEX
GMT-MASTER
SUPERLATIVE CHRONOMETER
OFFICIALLY CERTIFIED

GMT-MASTER
OFFICIALLY CERTIFIED
SWISS

At Baselworld 2014, as a world première, Rolex presented a new version of the Oyster Perpetual GMT-Master II with a bi-colour bezel in red and blue Cerachrom: a perfect blend of cutting-edge technology and the iconic aesthetics of the original 1955 model.

The GMT-Master with a case that is illustrated here (ref. 6542) was auctioned for CHF166,300 in 2007.

Opposite page
The first GMT-Master, ref. 6542, had a Bakelite bezel, which was later replaced by a metal one. The first bezels were red and blue, but black ones were soon introduced (ref. 1675).

LADY OYSTER PERPETUAL | 1954

The history of the Oyster commences with a woman, Mercedes Gleitze, who became the first testimonial of the waterproof case when swimming across the English Channel in 1927.

In 1931 the combination of the Oyster case and the Perpetual winding mechanism led to the manufacture and sale of a large and popular series of Oyster Perpetual models. It was 1954, however, before the first Lady Oyster Perpetual models were manufactured. Women's Oyster watches are now available with diameters of four different sizes: 26, 31, 34 and 36mm. Only the two larger versions have a calendar. These models are in steel or in a combination of steel and white gold, with pink, blue or silver dials.
The triple bracelet used for this collection is the Oyster, which is marked by smooth edges and an elegant yet simple design.
This is a sports collection designed for women.

In 1954 Rolex launched the women's version of the Oyster Perpetual: ref. 6509 – the first watch guaranteed waterproof to a depth of 50 metres. These models are currently available with a smooth (ref. 176200) or knurled bezel (ref. 176234).

Opposite page
A 1950s Rolex advert for the launching of the Lady Oyster Perpetual on the Italian market.

ROLEX

presenta l'OYSTER PERPETUAL per signora

il primo orologio del mondo impermeabile ed a carica automatica

BREVETTO MONDIALE ROLEX
DEL 1926
30 ANNI DI ESPERIENZA
UNA REPUTAZIONE MONDIALE

Nuovo calibro piatto nel suo genere, con "ROTOR" che carica nei due sensi.
Incastonato in 27 Rubini.
Cassa "OYSTER" perfettamente impermeabile garantita a 50 metri nel mare (5 atmosfere).
Anti-urto e antimagnetico.
Molla infrangibile.
Vetro infrangibile - non si scheggia.
Corona di carica "Twinlock" a sicurezza doppia.
Sospensione elastica del "Rotor" che lo protegge dalle vibrazioni; carica massima della molla eliminata grazie ad un dispositivo speciale.

Si fabbrica NEL MODELLO 27 RUBINI PRECISIONE oppure 27 RUBINI CRONOMETRO CON BOLLETTINO UFFICIALE SVIZZERO DI OSSERVAZIONE UN CAPOLAVORO TECNICO ROLEX.

MODELLO COME ILLUSTRAZIONE G. 6509

Cassa acciaio e oro, cuoio	L. 79.500
Cassa acciaio e oro con bracciale acciaio e oro estensibile	L. 100.000
Cassa oro 18 carati, cifre e sfere oro lùminose cuoio	L. 127.500
Cassa oro 18 carati, cifre e sfere oro luminose, bracciale oro 18 carati estensibile	L. 221.500

CON BOLLETTINO UFFICIALE DI OSSERVAZIONE Lit. 17.000 di SUPPLEMENTO NELLE DIVERSE ESECUZIONI.

IL PRINCIPE DEGLI OROLOGI

ROLEX

Montres ROLEX S. A.
GINEVRA

LA PIÙ FORTE PRODUZIONE MONDIALE DI CRONOMETRI
UFFICIALMENTE CONTROLLATI
Direttore Generale: H. Wilsdorf

MILANO

Ronchi Renato - Via Gonzaga, 5 (Piazza Diaz)
Verga Romolo - Piazza del Duomo, 19
Orologeria Giudici - Largo S. Margherita
Orologeria Luigi Verga - Via Mazzini già Via Carlo Alberto
Pisa Osvaldo - Via Pietro Verri, 9 (Angolo Montenapoleone)
Calderoni S.A. - Via Montenapoleone, 8
Orologeria Pisa - C.so Venezia, 5 (San Babila)

ROMA

Bandiera & Bedetti - Via del Teatro Marcello, 26
Cravanzola Succ. Gardino - Via del Corso, 341
Ditta Pelloni - Via del Corso, 140
G. Monetti - Via delle Convertite, 15
M. Bedetti & C. - Piazza S. Silvestro, 11

Alessandria - Maccarini-Menghi - Piazza Lega Lombarda, 2
Bari - Calderoni S.A. - Via Vitt. Veneto, 36
Benevento - Cangiano Aldo - C. Garibaldi, 175
Bergamo - G. Curnis - Via Monte Grappa, 7 (Portici Sentierone)
Biella - F. Albonico - Via Italia, 11
Bologna - F. Veronesi & Figli - P.za Mag. 4
Bolzano - G. Pörnbacher - Portici, 51
Brescia - M. Schreiber - Corso Zanardelli, 17
Busto Arsizio - G. Mazzucchelli - Piazza San Giovanni, 1
Cagliari: Casa del Gioiello - C. Vitt. Em. 63
Catania - G. Avolio & Figlio - Via Etnea, 15
Chianciano - G. Torrini & Figli - Gioielleria-Orologeria
Como - M. Dell'Oca - C.so Vitt. Emanuele
Cortina d'Ampezzo - Punzo Vincenzo - Corso Italia, 8
Cremona: Brugnelli Ottorino - Via Verdi, 6
Firenze - Orologeria Svizzera - Piazza San Giovanni, 3
Foggia - Dante Ponzio - Via Lanza, 4
Formia - Giuseppe Mirante - Via Vitruvio, 36
Gallarate - Alfredo Ganosa - Gioielleria
Genova - F. Chiappe - Piazza Dante, 9/4
Genova - Luigi Codevilla - Via Orefici, 55
Genova - Montres & Bijoux S.A. - Piazza De Ferrari, 3 r
Genova - Pavese Cristoforo - Via Roma, 24 r
La Spezia: Giorgini Ilario - P. Mentana, 20 r
Livorno - Caracciolo - Corso Amedeo, 16
Messina - F.lli La Motta - Viale San Martino, 188
Modena - F.lli Blondi - Via Emilia, 90
Monfalcone - A. Cernigoi - Via Friuli, 62
Montecatini Terme - V. Punzo - Portici, 4 A
Monza - Zaccaria Viganoni - Via Italia, 16
Mortara - Carlo Baiardi - Corso Garibaldi, 1
Napoli - G. Monetti - Via S. Brigida, 60 A
Novara - Luigi Baiardi - Via Gramsci, 15
Padova - Ermanno Bergamo & C. - Via Cavour, 3
Padova - Gioielleria Smeraldo - P.za Frutta, 8
Palermo - A. Matranga - Via Maqueda, 274
Parma - S. Verdoni - Corso Repubblica, 4
Pavia - Alessandro Redaelli - Strada Nuova, 68
Perugia - A. Biagini - Via G. Calderini, 13
Piacenza: E. Della Lucia - C. Vitt. Em, 51
Potenza - A. Tomasco - Via F.lli Assisi, 10
Prato - Ditta Gironi - Meoni Dino - Via Ricasoli, 5 R
Ragusa: G. Schininà-Calabrò - V. Roma, 129
Rapallo - Nando Maddalozzo - Via Cairoli, 2
Reggio Calabria - T. Versace - V.le Garibaldi
Riccione: G. Tamburini - V.le Ceccarin, 19
Rimini - E Tamburini - P.za Tre Martiri, 70
Salerno - Antonio Galdi - Corso Vittorio Emanuele, 111
San Remo - M. Abate - Corso Imperatrice, 1
Savona - A. Dupanloup - Piazza Mameli, 4
Taurianova - T. Versace - Gioielliere
Terni - F. Festuccia - C.so Tacito, 81
Torino - V. Astrua - Via Roma, 6
Trento - Luigi Tapparelli - Via Oriola, 31
Trieste - G. Dobner - Via Dante, 7
Udine - I. Ronzoni - Via Mercatovecchio, 10
Udine - Santi di I. Ronzoni - Via del Monte, 6
Venezia - G. Salvadori - Mercerie S. Salvatore, 5022
Venezia - L. Benvenuti - San Marco, 262
Vercelli - Enrico Natti - Corso Libertà, 14
Verona - A. Canestrari - Via Cappello, 35 (Via Mazzini)
Vigevano - Ermanno Gian Rossi - Via del Popolo, 28
Mogadiscio (Somalia) - Lugli & Zini - Orologeria - Via Tommaso Duca di Genova

DAY-DATE | 1956

At its launch at the Basel Fair in 1956, the Day-Date was showcased as a world novelty: it was the first watch to indicate the day of the week spelled out in full.

Other innovations of the original model include a balance with the Microstella regulating screw and a mechanical alteration that made it the first Rolex watch to be classified as "Superlative Chronometer Officially Certified" for accuracy. In 1977, the plastic glass was replaced with a sapphire crystal glass, making it waterproof up to 100 metres.

Another important change was made in 1988 with the introduction of the independent regulation of the date and day of the week.

Present-day models can display the days in any language and are almost invariably fitted with the President bracelet. This takes its name from US President Dwight D. Eisenhower, who was given one of the first pieces from this series as a gift. The President bracelet has an invisible clasp and was especially designed for the Day-Date. This watch is available in various shades of gold or platinum. In 2012 Rolex launched some models with a chocolate-colour dial studded with rubies and diamonds.

Ref. 118235, the Everose pink gold version, which was introduced in 2012 with a chocolate-colour dial, eight brilliant-cut diamonds as hour-markers and two baguette-cut rubies at 6 and 9 o'clock. The calendar is set at 12 o'clock, the Cyclops aperture at 3.

MONDAY
ROLEX
OYSTER PERPETUAL
DAY-DATE
28
SUPERLATIVE CHRONOMETER
OFFICIALLY CERTIFIED
SWISS
MADE

MONDAY
ROLEX
14
DAY-DATE

In 1961 ref. 1803 made its first appearance in Rolex catalogues. It remained in production until 1977-1978. The model illustrated here is in white gold with a knurled bezel and Plexiglas face. In 1977 Rolex launched the Oysterquartz Day-Date; ref. 19019 is in white gold.

Opposite page
Developed in the glamorous 1980s, the white gold ref. 19049 has a bezel set with 44 diamonds and a diamond-studded Pleiade dial.

MONDAY
ROLEX
28

MONDAY
ROLEX
28

By the 1980s, the Day-Date was being embellished through the addition of diamonds with different cuts on its crown.

Opposite page
Ref. 118238 from the Day-Date collection and Ref. 218398 from the Day-Date II collection are featured here with dials adorned with baguette-cut rubies, baguette-cut diamonds, and brilliants.

MILGAUSS | 1956

First presented in 1956, the Milgauss – the name is a combination of the French word 'mille' (thousand) and the unit of magnetic induction, 'gauss' – was designed in such a way as to ensure the perfect functioning of the watch even in the presence of strong magnetic fields.

As certified by the CERN (European Organization for Nuclear Research), the watch has an endurance of up to 1,000 gauss (the average figure being 100). This is thanks to a soft-iron grid that protects the calibre from external electromagnetic fields; also, the movement is made of an anti-magnetic alloy.
Manufactured throughout the 1960s and '70s, the Milgauss vanished from Rolex catalogues circa 1990.
It was newly showcased at the Basel Fair in 2007 with reference 116400: this model preserves the anti-magnetic qualities of the original version, adding new features such as the 40mm case, a polished finish and a white or black dial. All Milgauss watches are set with a COSC-certified self-winding mechanical movement.
Nowadays, the Milgauss is furnished with a Parachrom hairspring in an innovative anti-magnetic material, to ensure further protection against magnetic forces.

Ref. 1019 was in production between 1966 and 1990.

Opposite page
The 1956 Ref. 6541 with steel case and bracelet, black dial and ring.

Presented in 2007, ref. 116400 has a 40mm carrure, a polished finish and a black or white dial, shielded by a green sapphire crystal.

ROLEX
OYSTER-PERPETUAL

20
10

Rolex ads from the mid-1950s.

Opposite page
Ref. 6541 is known as "saettone" in Italian because of its second hand in the shape of an arrow – "saetta" in Italian. Resistance to magnetic force was ensured not only by the use of anti-magnetic material, but also by a second soft-iron back, fitted precisely inside the case, so as to seal it against magnetic forces.

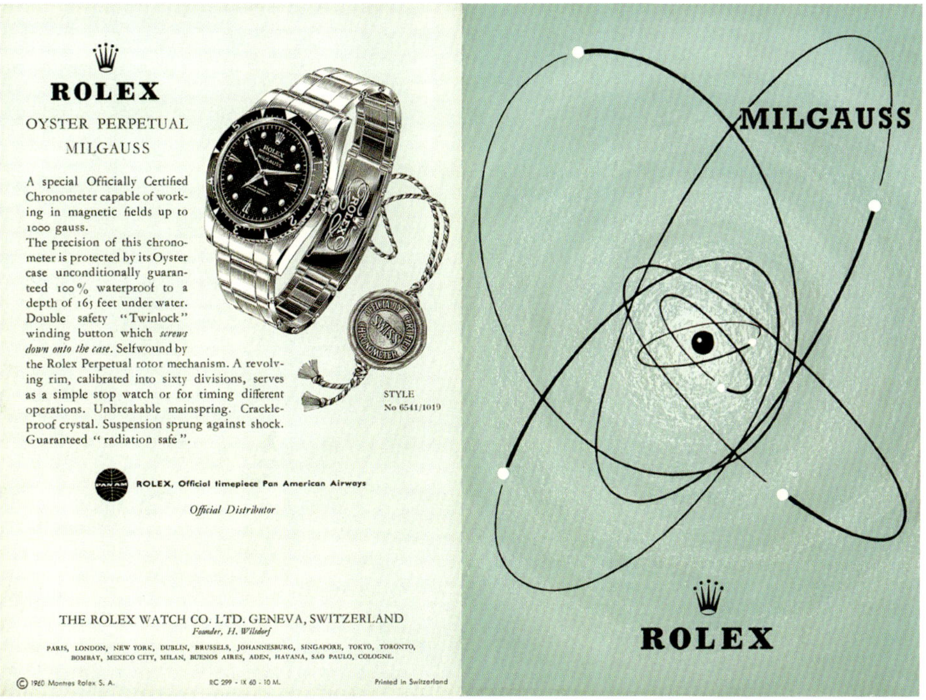

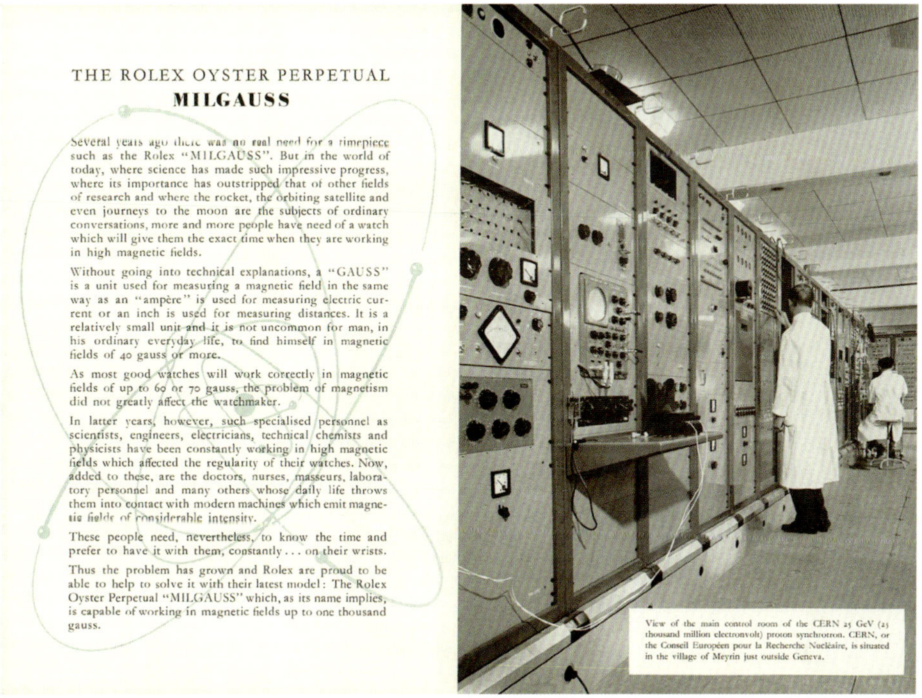

AIR KING | 1957

During the Second World War, in order to meet the needs of servicemen and aviators, the Rolex maison *launched a series of Oyster watches with a case diameter of 33mm – a slightly larger model compared to the conventional 32/33mm one.*

These are resistant and luminous watches with an easy-to-read dial.

These models, which were particularly favoured by British RAF pilots, also proved quite popular with the wider public. The Air King model continued to be marketed after the war without ever falling out of production. The Air King, which is not certified as a chronometer, is the most affordable Rolex. Many customised versions of this model have been ordered.

In the "special series" with a customised dial, the word "Precision" is replaced by the name of the company that has ordered the watch, though the name "Air King" is retained – with its distinctive rounded lines for the letters "A" and "K".

Introduced in 1957, ref. 5500 remained on sale until 1990. Versions with a customised dial are very common. The model featured here is a Win Dixie.
Ref. 14010 was introduced in 1996.

ROLEX
GENEVA
ROLEX
OYSTER PERPETUAL
Air-King
PRECISION
SWISS T
ROLEX
OYSTER PERPETUAL
Air-King
PRECISION
T SWISS MADE T

DAYTONA | 1963

This hand-wound chronograph from the early 1960s was originally launched as the Rolex Cosmograph (the name Daytona was only added later), and was given a rather lukewarm reception.

At the time, more classic-looking, ultra-thin watches were in vogue: a model as sporty and modern as this was not to most people's taste.
The first models were hand-wound: only in 1988 did Rolex introduce the Zenith El Primero movement, the best self-winding movement then available on the market, but with several alterations to bring it up to the company's standards.
In 1988, the Oyster Perpetual Cosmograph Daytona was favourably received by the public. During the early 1990s, the limited production of this model, partly due to the fact that the basic Zenith calibre was not widely available, excited collectors' zeal. In the year 2000, Rolex presented the new Daytona ref. 116520, a model manufactured entirely by Rolex. The engraved bezel of the Cosmograph Daytona can be used as a tachometer to measure a speed of up to 400 units per hour and record the time-lapse from a given distance – both in miles and kilometres. The central hand of the chronograph is accurate to 1/8 of a second. In present-day models, the two subdials set at 3 and 9 o'clock record the hours and minutes lapsed, enabling the wearer to plan the time for the laps in a race.
In 2012, Rolex launched a model with a ring set with coloured gems that formed a rainbow around the dial.

Ref. 6269 (hand-wound) has a ring studded with 44 brilliant-cut diamonds.

Opposite page
Many different versions of the Daytona Perpetual have been developed, including limited editions distinguished by their colour or by the finish of their dial. Ref. 16518 has a white dial with white and champagne chronograph subdials (ebauche Zenith); Ref. 16519 features a pearl white dial; Ref. 116519, a chrysoprase green dial (with a Rolex movement).

Ten different models with a blue dial have been produced for ref. 16528. Ref. 6265 (hand-wound) is a one-off featuring a red and black dial.

The dial of ref. 6241 (hand-wound) plays on the contrast between white and red.

Right
A detail of the dial with diamond-studded face.

DAYTONA REFERENCES

References and manufacture dates for the hand-wound Cosmograph Daytona. The references are listed in numerical order according to type:

6239 Produced between 1963 and 1969; pump pushers; metal ring; 72B or 722 calibre movement at first, then 722-1.

6241 Produced between 1966 and 1969; pump pushers; metal/plastic ring; 72B or 722 calibre movement at first, then 722-1.

6262 Only produced in 1970; pump pushers; metal ring; 727 calibre movement.

6264 Produced between 1970 and 1972; pump pushers; metal/plastic ring; 727 calibre movement.

6240 Produced between 1965 and 1969; screw-down pushers known as "millerighe" ("thousand lines"); metal/plastic ring (2nd and 3rd series); 72, 72B or 722 calibre movement at first, then 722-1.

6263 Produced between 1971 and 1988; screw-down pushers (2nd and 3rd series); metal/plastic ring; 727 calibre movement.

6265 Produced between 1971 and 1988; screw-down pushers (2nd and 3rd series); metal ring; 727 calibre movement.

6269 Produced between 1984 and 1988; screw-down pushers (2nd and 3rd series); gold and 44-diamond ring; 727 calibre movement.

6270 Produced between 1984 and 1988; screw-down pushers (3rd series); gold ring with 28 emerald-cut diamonds; 727 calibre movement.

In 2014, Rolex introduced a new jewel edition of the Oyster Perpetual Cosmograph Daytona, combining the nobility of 950 platinum with the brightness of the purest diamonds. The entire dial is pavé-set with diamonds, with the exception of the chronograph counters, which are ice blue: a rare and exclusive colour that Rolex only uses for platinum models. The bezel is highlighted by the radiance of 36 baguette-cut diamonds.

Ref. 16520 features a vary rare colour-change black dial with brown chronograph counters, known as the "Patrizzi". Ref. 6262, produced in 1970, is marked by a black dial with white sub-dials.

SEA-DWELLER | 1967

The Sea-Dweller springs from the collaboration between Rolex and Comex (Compagnie Marittime d'Expertise), a French company specialising in underwater engineering.

Initially the divers working for this French company were using the Submariner, which was waterproof up to 200 metres, but there was a risk that the pressure from the helium in the diving bells would shatter the glass of the watch. Rolex came up with a solution to meet these specific requirements. Some Submariner models started featuring a small valve opposite the winding crown, consisting of a unidirectional spring-loaded piston sealed by a gasket. When the watch is plunged to a certain depth, the pressure of the water pushes the piston against the gasket, making the watch waterproof; during the resurfacing phase, the pressure exercised by the helium inside the watch opens the valve, allowing the gas to escape. The first Sea-Dweller made its appearance in 1967. During the 1980s, a special kit was added with a tool to remove the links and the spring lugs, an extension for the bracelet and a table indicating decompression times.

In 2008, almost 55 years after the first feat accomplished with the Deep Sea Special model, Rolex released a new issue of the Sea-Dweller: waterproof to 3,900 metres.

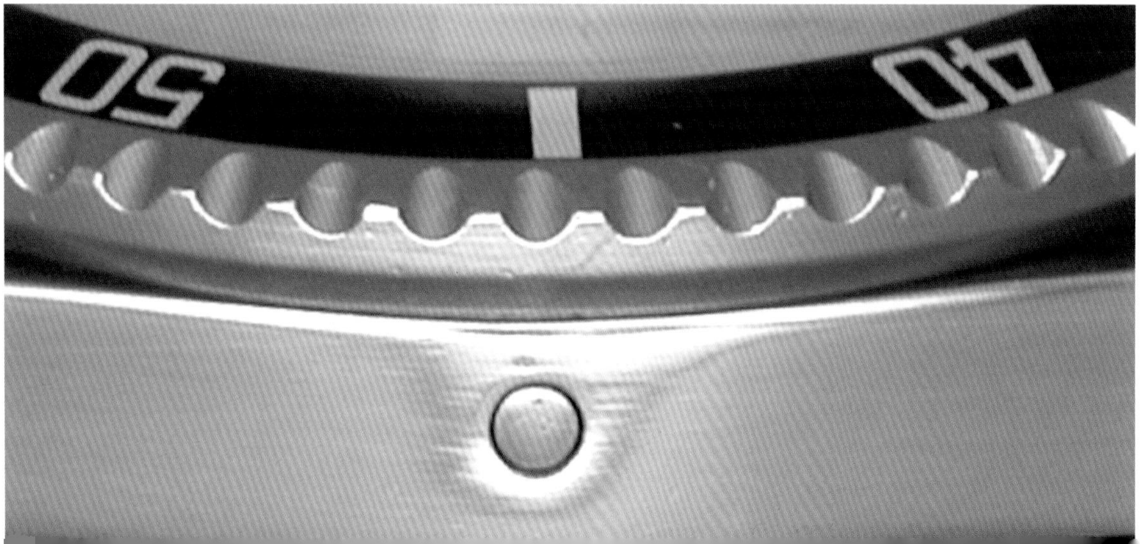

In 1969, Rolex produced a steel watch with a valve located in the case and positioned at 9 o'clock, allowing helium and other gases to escape, thus guaranteeing the protection of the movement during the decompression stage (ref. 1665).

ROLEX
PERPETUAL DATE
SEA-DWELLER
SUBMARINER
SUPERLATIVE CHRONOMETER
OFFICIALLY CERTIFIED
SWISS-T<25

ROLEX
SEA-DWELLER

ROLEX
OYSTER PERPETUAL DATE
SWISS - T<25

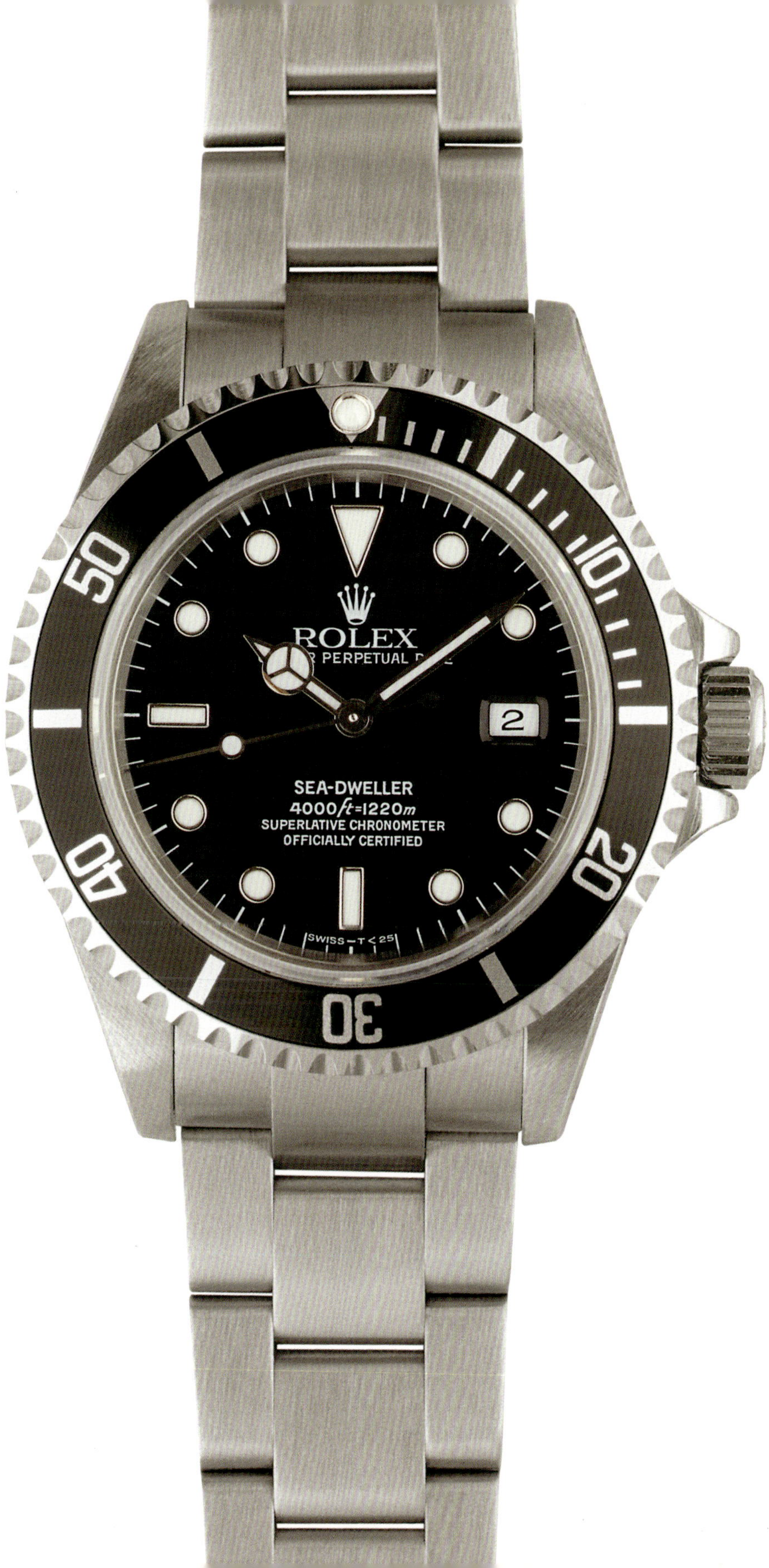

Ref. 16600, launched in 1986, is waterproof to 1,220m. Its sapphire glass is 1mm thicker than that of the Submariner, enabling it to withstand a greater depth.

LADY DATEJUST | 1965

In 1965, the first Lady Datejust was introduced: ref. 6517. However, only the red-gold Rolesor version was actually named Datejust.

On the Rolesor Yellow, Rolesor White, yellow gold and white gold versions of this reference, the only word that appears on the dial is "Date". Between 1970 and 2006 Rolex manufactured a Lady Datejust - until 1983 with a 2035 movement and thereafter with a 2135 movement - without the quick-set date function.

In 1983, Rolex started producing both the Day Date version (ref. 69240) and the Datejust version (ref. 69163) of the Lady model, both running on a 2135 movement.

Today, the Lady Datejust collection includes a wide selection of models, with a case diameter ranging from 26mm in the steel, steel/white gold, steel/yellow gold, steel/Everose gold, Everose gold and platinum versions. The dials have a guilloché finish or come in the Gold Crystals version, created through a process of crystallisation of gold, and available in grey, pink or yellow. Some models have diamonds for hands, or a diamond-studded bezel.

The Lady Datejust Pearlmaster models have a case with a 29mm diameter and are marked by softer, more rounded lines compared to the classic model.

Available in yellow gold (ref. 179178) or white gold (ref. 179159), the new models from the Lady Datejust series have Roman numerals, with the number 'VI' formed by 19 gems - rubies on the gold model or sapphires on the white gold model. Ref. 179175 has a diamond-studded pearl dial; ref. 80309 (Pearlmaster), a bezel with baguette-cut sapphires and a bracelet set with diamonds and sapphires.

ROLEX
OYSTER PERPETUAL
DATEJUST

YACHT-MASTER | 1992

In 1992, Rolex presented the Yacht-Master model, devoted to the sailing world – especially yachtsmen and skippers.

This yellow-gold watch has a screw-down, 40mm, three-body case, a unidirectional rotating bezel, a sapphire crystal, a Triplock screw-down crown and an Oysterlock bracelet.
The second-generation Yacht-Master, known as Yacht-Master II, made its début on the watch market at the Basel World Fair in 2007. It is a technologically highly innovative model. The Yacht-Master II is the first Rolex to have a 10-minute countdown graduation with a mechanical memory. It provides precise synchronisation with the starting sequence of a yacht race. The new models have a polished case, clean-cut lugs, a Triplock winding crown with protective guards and two start/stop pushers set at 2 o'clock to start and stop the countdown, and a reset pusher at 4 o'clock to programme and synchronise the countdown. The white dial with luminous square hour-markers, red seconds subdial at the centre and stopwatch hand - with a red-bordered white triangle at the end - ensure optimum legibility; the Oyster bracelet has an Oysterlock clasp and the Easylink extension, which makes it comfortable to wear with any clothing - even over oilskins.

A version of the Yacht-Master made of Rolesium (a steel and platinum alloy) was presented at the Basel Fair in 2012. Ref. 116622 has a blue dial.

ROLEX
SWISS MADE

Ref. 16628 is a yellow-gold model with a diameter of 40mm. It has a rotating graduated bezel and a case with protective guards for the winding crown.

Opposite page
In 1994, a 29mm women's version was introduced (ref. 69628), along with a 35mm men's version (ref. 68628).

ROLEX
28

ROLEX
28
ROLEX
28

SKY-DWELLER | 2012

The Oyster Perpetual Sky-Dweller, presented at Baselworld 2012, is the latest addition to the Oyster range.

With its fourteen patents, including five new deposits, the Sky-Dweller can provide all the information grand voyagers need to find their bearings. The dual time-zone with the local time is indicated by the central hands, while the reference time is displayed by the off-centre rotating disk on the dial. In addition to this, there is the annual calendar Saros – which takes its name from the astronomical phenomenon of the cyclical alignment of Sun, Earth and Moon, and which requires re-setting only once a year, between February and March – and the indication of the months within twelve apertures along the edge of the dial. To adjust the local time, reference time and calendar quickly, the watch makes use of an extremely innovative interface between the case and the movement. The focal point of this interface is the rotatable Ring Command bezel, which allows the wearer to select the individual functions to be set simply by turning the bezel. Each function can then be adjusted swiftly in both directions by means of the winding crown, which has only one setting position. The Sky-Dweller is equipped with the new 9001 calibre, officially certified as a chronometer.

Sky-Dweller ref. 326935 is an Everose model with knurled bezel, brown dial and alligator strap. Ref. 326938 is a yellow-gold model with a bracelet made of the same material.

Opposite page
Ref. 326939, in white gold.

ROLEX
SKY-DWELLER
SWISS
MADE

ROLEX
OYSTER CHRONOGRAPH
ANTI-MAGNETIC
GIO
AGO

RECORD VINTAGE PIECES

REF. 3525

CHF327,000 - GENEVA, 17 NOVEMBER 2008

Reference 3525 bears witness to the human qualities of Hans Wilsdorf, who was not only a great entrepreneur, but also a socially engaged man.

During the Second World War, the Nazis would seize the personal effects of all soldiers held captive, including their watches. Wilsdorf struck an agreement with the International Red Cross allowing his representatives to give a watch to any British officer who might be interested. Upon their release, the officers could choose whether to buy the watch at a reduced price or to return it to Rolex with no further obligations. Approximately 200 copies of the steel chronograph Reference 3525, and nicknamed Prisoner's Watch, were produced between 1940 and 1943. One such 3525 model was sold at a Christie's auction in Geneva on 17 November 2008 for 327,000 Swiss Francs. This watch was part of the Mondani collection, which had been put on the auction block by Antiquorum/Patrizzi in Geneva in 2006. It is one of the best-preserved copies of the famous chronograph. Manufactured in 1941, it belongs to the first series of this model; it is marked with a reference and series number on the outer part of the back, rather than between the lugs as on later models. It has a black enamel dial with a gold pattern and shows a pulsometric scale rather than the telemetric and tachometric scales found on the standard version - a detail that makes it all the more rare and valuable. All this suggests that the watch was designed for a medical officer.

35mm mono-block case, screw-down bottom and crown, black dial with gold pattern. "Valijux 13" movement, pump push-pieces, two counters – one displaying the constant seconds, the other the minutes – and pulsometric scale. Signed case, dial and movement.

REF. 4113

CHF1,035,000 - GENEVA, 16 MAY 2011

This split-second chronograph – ref. 4113 and case no. 051318 – was the first to cross the CHF1 million mark. It is one of the twelve reference 4113 watches produced in 1942.

Since this model was never made available to the public, nor illustrated or referred to in Rolex ads from the period, its origins have long been a matter of speculation. The research conducted so far has yielded contradictory results.

One investigation conducted by Christie's suggests a direct link with the world of automobile racing. The first 4113 ever put on the market, which was sold in Geneva on 15 May 1991, had formerly been owned by a relative of Stefano La Motta, Baron of Salinella, a famous Sicilian pilot. According to other sources, since car races - including the Targa Florio, held in Sicily - were suspended in 1941 and only resumed in 1948, these watches cannot have been designed for racers, but were probably commissioned by the Italian Rolex supplier for the Italian aircraft, without ever being delivered. At the end of the war the twelve watches still in stock must have been marketed and sold as split-second chronographs to the top pilots of the time.

Aside from Sicily, 4113 chronographs also circulated in England. The watch featured here had already been auctioned by Christie's in 1991, when it was owned by the widow of a member of a motor racing team. These rare pieces are highly sought-after by collectors.

Chronograph running on a 55UBR calibre movement manufactured by Valjoux, with 17 lines and 3/4, mono-metallic balance with screws, Breguet spring, rhodium finish, split-second hand controlled by a push-piece coaxial to the crown. The mechanism of the split-second chronograph is controlled by two rectangular push-pieces at the sides of the crown and by a round push-piece on the crown. Signed case, dial and movement.

REF. 4113

CHF1,107,000 – GENEVA, 13 MAY 2013

The split-second chronograph was mostly used to measure times in sporting competitions before the spread of digital stopwatches.

To ensure a double time measurement, it has two superimposed hands that set off simultaneously but can be stopped at separate moments. The mechanical device behind these functions is a very delicate one requiring a high manufacturing cost. Rolex only ever manufactured twelve split-second chronographs with reference 4113; they date from 1942 and have case numbers ranging from 051313 to 051324.

Over the past 30 years, only five of these watches have been put to auction: the model is held in such high regard that owners are seldom willing to part with them. Whenever these chronographs have been put on sale, they have always fetched remarkable prices. Indeed, it was a 4113 model that was the first Rolex watch to cross the CHF1million mark in 2011 (see p106). This record was superseded by another split-second chronograph from the same series in 2013, reaching the princely sum of 1,107,750 Swiss Francs.

The high price of this watch (ref. 4113, case no. 051314) is due to its fine state of conservation and its rarity. It had previously been auctioned by Christie's in New York in 1996, attracting the interest of many international bidders and ultimately fetching a price four times its pre-sale estimate. Almost 20 years on, this chronograph has once again caused a real stir on the market.

This is an outstanding split-second chronograph on account both of its aesthetics and its size. The case is in steel, the matte silver dial has Arab numerals and applied baton hour-markers in pink gold. The carrure, with a diameter of 44mm, and the very thin bezel increase the legibility of the dial, which has the constant seconds counter at 9 o'clock and the 30-minutes counter at 3. The telemetric scale is blue, the tachometric scale is black. Signed case, dial and movement.

17 RUBIS
ROLEX GENEVE

REF. 6036

$614,000 – NEW YORK, 14 DECEMBER 2012

Rolex has generally avoided producing watches with complications. One exception is those models equipped with a complete calendar (date, day and month) and a chronograph function, such as the famous "Dato-compax".

This model owes its name to the fact that it has a complete date calendar (Dato) and three subsidiary chronograph dials, also known as "Compax". A complicated watch, known as "Killy", it was created in 1947. Over the years, it was only marketed in small numbers with references 4767, 4768, 5036, 6036 and 6236.

The watch with ref. 6036 and case no. 113.762 was manufactured in 1955. In a perfect state of conservation, it was sold at a Christie's auction in New York for $614,000. The day and month indications in Italian and the original guarantee stamped with the name of the Rolex retailer Enrico Casalini of Novara confirm that this watch was first sold on the Italian market.

Another example of this reference – with case no. 113.760, which suggests a date of manufacture very close to that of the first watch – was sold at a Christie's auction in Geneva in 2010 for CHF440,000. This watch also features time indications in Italian and was probably a gift from the Italian National Olympic Committee.

Triple calendar chronograph with three-body tonneau-shaped case in pink gold with a diameter of 36mm. Bicolour silver dial with Roman numerals and applied dot hour-markers in pink gold. Luminous hands, external date ring with blue Arab numerals, central arrow date hand, day and month apertures at 12 o'clock with indications in Italian, three subsidiary dials for constant seconds, 30 minutes and 12 hours registers with flinqué finish. Signed case, dial and movement.

GIO
AGO
ROLEX
ANTI-MAGNETIC

REF. 6236

$638,500 - NEW YORK, 14 DECEMBER 2012

This watch is named after the famous French skier Jean-Claude Killy, who in the 1960s won three Olympic gold medals and two Alpine Skiing World Cups, and was very fond of this model – Rolex ref. 6263.

In November 2012 this watch set a new world record for its reference when, from a starting price of $120-180,000 dollars, it ultimately reached $638,500.

The Killy chronographs, with a triple calendar, are the most complicated watches ever created by Rolex. They remained in production for twenty-odd years from 1940; in steel, yellow gold and pink gold, they include references 4767, 5036, 6036 and 6236. The last three references are distinguished by a three-body, rather than two-body, case and by a larger and more prominent bezel.

The steel watch featured here has a bicolour dial. This example of reference 6236 comes from the collection of Gordon Bethune, the former CEO of Continental Airlines and an avid watch collector.

The perfect state of conservation of the case and dial, along with the original bracelet, fuelled competition among collectors, leading to a record result.

Anti-magnetic Oyster Chronograph watch from 1960, ref. 6236, case no. 576.392. Case with a 36mm diameter and steel bracelet. Matte silver dial, spear-shaped hour-markers and applied luminous numerals. Hands with luminous inserts, outer date ring with blue Arab numerals, arrow date hand at the centre, day and month apertures at 12 o'clock, three subsidiary dials for constant seconds, 30 minutes and 12 hours registers with flinqué finish. Signed case, dial and movement.

REF. 6062

$590,500 – NEW YORK, 14 DECEMBER 2012

For the first time in its history, between 1950 and 1953 the maison sought to equip an Oyster cased watch with a moon-phase complication.

Reference 6062 was manufactured in steel and yellow or pink gold. It is estimated that no more than 500 copies were produced.

The gold versions feature star-shaped hour-markers on some dials, which earned the model the nickname of "Stelline" (Little Stars).

On 18 January 1992, at the Antiquorum/Patrizzi "The Art of Rolex" auction held in Vicenza, a "Stelline" watch was put on the block for 50 million lire. At the Christie's auction held in Geneva on 16 May 2005, a 6062 fetched CHF356,000; then on 14 December 2012, at another Christie's auction, a pink-gold version of this model from 1950 was sold for $590,000. This watch from the collection of Gordon Bethune, the former CEO of Continental Airlines, is in perfect condition. The dial retains its subtle *grainé* finish, along with the original printing. The luminous hour markers and the luminous material on the hands have, with time, taken on a warm and charming mustard tone. The case shows a wonderful dark patina from oxidation and the back retains its satin finish. Finally, this example still has its original Super-Oyster pressure crown.

Moon-phase watch with 36mm pink-gold case. The matte silver dial has star-shaped hour-markers applied to the outer perimeter. The hands have a dauphine design, with a luminous insert at the centre. The arrow-shaped date hand has a blue coating. The apertures displaying the day and month are set at 12 o'clock. The moon phase indication is set at 6. The watch runs on an automatic 655 calibre movement with 18 rubies. Signed case, dial and movement.

Christie's sold a yellow gold 6062 model in New York in December 2011 for $542,500.

REF. 8171

$1,145,000 – NEW YORK, 17 DECEMBER 2013

This model was given the nickname "Padellone" (Big Pan) by Italian collectors on account of its rather large case (with a diameter of 38mm). It is now come to be known by this name around the world.

Rolex has only created two wristwatch models with a moon-phase function: references 8171 and 6062. The former has a push-back case, the latter an Oyster case. Reference 8171 was in production roughly between 1949 and 1952, and it is estimated that no more than 1,000 copies were ever manufactured by the *maison* – in yellow gold, or more rarely in steel and pink gold. When this watch, which is in remarkable condition, was first put on the auction block by Christie's in New York in December 2013, collectors vied to outbid one another: the 'Padellone' was ultimately sold for $1,145,000, four times the starting price.

This 8171 from 1950 is probably one of the finest watches of its kind in terms of its state of conservation. The design of its dial presents a number of features that make it unique, such as the crown logo at the top and the inscription "Perpetual" under the date apertures.

After its record sale, the watch earned the new nickname of "Sleeping Beauty" because it came as a real surprise to buyers.

Anti-magnetic and astronomical automatic watch with triple date and moon phases. The pink-gold push-back case has a diameter of 38mm. The bicolour silver dial has leaf-shaped hands and a second hand in burnished steel with a blue arrow-shaped tip. The seconds subdial, set at 6 o'clock, encloses a blue aperture for the moon phases adorned with golden stars and moon. In the 12 o'clock position, just below the Rolex crown, are an aperture with the day of the week on the left and one with the month on the right. Signed case, dial and movement.

REF. 6239

CHF761,000 – GENEVA, 10 NOVEMBER 2013

A model with highly specific functions may not be appreciated by the public when it is first launched on the market, in which case it will only be manufactured in a limited number of copies.

Years later, the model will be highly sought after by collectors because of its rarity. This is the case with the Daytona watch with a pulsometric scale, only ten copies of which are known. The pulsometric scale is a function designed for doctors. It enables one to measure the pulse rate by starting the chronograph function at the first beat and stopping it at the fifteenth, without having to measure the beats for a whole minute. The number of beats per minute is indicated by the chrono hand on the pulsometric scale (up to 200 beats). The addition of the pulsometric scale led Rolex to make some changes to the dial. The subdials are slightly larger and the colour scheme is also different from that of other models: the medical scale is bright blue, ensuring perfect legibility despite the small size of the numerals. The medical Daytona, or so-called "Doctor Watch", is no doubt one of the rarest chronographs; and of the known "Doctor Watches", only two have the name "Daytona" in red. This one is the Rolex Cosmograph Daytona ref. 6239 from 1967, which was auctioned by Christie's for 761,000 Swiss Francs.

Chronograph with 37mm tonneau-shaped case and steel Oyster bracelet. The burnished silver dial has applied baton hour-markers with luminous inserts, hands with luminous inserts, and a pulsometric scale calibrated to measure the heart beat. The three subsidiary dials for constant seconds, 30-minute and 12-hour registers are in black. The white ring is graduated to 300 units per hour. The watch has a screw-down back and crown, and runs on a 722-1 calibre mechanical movement with 17 rubies. Signed case, dial and movement.

REF. 6263 - PAUL NEWMAN

CHF989,000 - GENEVA, 10 NOVEMBER 2013

Around 1963 Rolex started marketing a Daytona watch with a bicolour dial known as the "exotic" or "tropical" model.

Italian collectors gave this dial the new nickname of "Paul Newman" after it was worn by the famous actor in the poster for the film *Winning*, in which he stars in the leading role. The Paul Newman model is marked by the chromatic contrast between the dial and the counters, which have the same colour as the 60-minute scale. The dial was initially flat, but was later developed on two levels, with a small step set against the back: a design that increases the three-dimensional effect and leaves more room for the hands. Some specific features further increase the value of the watch featured here, which was auctioned by Christie's in November 2013 for 989,000 Swiss Francs. Among all the different versions of the Daytona Paul Newman, that with a black dial and screw-down push-pieces has become the most valuable and the most sought-after by collectors around the world. A rough estimate suggests that just over a dozen of these watches were ever released on the market.

37mm steel case with screw-down back and crown. 727 calibre mechanical movement with 17 rubies, black dial, square hour-markers with luminous inserts, luminous hands, three subsidiary dials for constant seconds, 30-minute and 12-hour registers with flinqué finish, and Oyster bracelet. Signed case, quadrant and movement.

130
140
150
160
180
200

REF. 6263 - PAUL NEWMAN

CHF841,300 – GENEVA, 12 MAY 2013

The year 2013 is one to be remembered in the history of Rolex, since the Antiquorum auction that was held in Geneva in May registered a remarkable sale.

The piece on the auction block was a ref. 6263 watch, a yellow gold Paul Newman Rolex Daytona featuring a rare lemon-yellow dial with contrasting black counters and 60-minute scale with white graphics.

This watch, manufactured in the late 1960s, was purchased – after a real battle for the highest bid – for 841,300 Swiss Francs, almost three times the starting price.

This is one of the rarest Daytona models ever marketed. Before the 2013 auction, it was unknown among collectors. This reference 6263 is marked with a very low series number, which is actually antecedent to that of some older models, such as references 6262 and 6264 (the last models with pump push-pieces).

The white graphics of the subdials and the ring, which create a pleasant contrast with their black background, increases the legibility of the watch. With time, the dial has acquired a far brighter yellow hue than the cream colour that is typically seen on these models.

The three-body 37mm dial is in yellow gold, with screw-down push-pieces. The ring is graduated to 200 units per hour. The watch has a bicolour yellow and black dial with square hour-markers in applied gold with luminous inserts, and three subsidiary dials for constant seconds, 30-minute and 12-hour registers. The baton hour-markers are in yellow gold with luminous inserts. The watch is equipped with a 727 calibre movement with 17 rubies, an anchor escapement, a mono-metallic balance with screws, a Breguet spring, and Microstella screws. Signed case, dial and movement.

ROLEX
OYSTER
COSMOGRAPH
UNITS PER HOUR
T SWISS T

REF. 6263 – CUSTOMISED WATCH

CHF785,000 – GENEVA, 10 NOVEMBER 2013

The reference 6263 Rolex Daytona personalised with the signature of Sultan Qaboos Bin Said of Oman far exceeded the sale estimates, which ranged between CHF100,000 and 300,000.

The watch was sold by Christie's in November 2013 for 785,000 Swiss francs.

In 1970, Qaboos Bin Said Al Said seized control of the small country of Oman and set out to strengthen its military, as well as commercial and diplomatic capacity, with the aim of earning international acknowledgement. As a gift for those loyal to him, the watch enthusiast Sultan commissioned several models from Rolex with a customised dial bearing either his own signature or the kingdom's emblem: two crossed swords superimposed on a khanjar, the typical Omani dagger. Manufactured in 1974 and sold by Asprey, the famous luxury boutique in London, which was the official Rolex distributor for the Middle East at the time, this Daytona bears His Majesty's signature in red at 12 o'clock. The graphics on the dial are also customised: the silver "Rolex Oyster" mark is set higher up than usual, while the word "Cosmograph" is in the 6 o'clock position. The watch was sold with its original box, also personalised with the emblem of the Sultanate and including a note from the Sultan himself, which reads "With My Compliments, Qaboos Bin Said, Sultan Of Oman".

Rolex Oyster Cosmograph with 37mm steel case. The black dial is personalised with the Sultan's signature in red. Applied silver baton hour-markers with luminous inserts and baton hands with luminous inserts. The three subsidiary dials are in silver with a flinqué finish, the black ring is graduated to 200 units per hour. The watch runs on a 727 calibre mechanical movement with 17 rubies, anchor escapement, mono-metallic balance with screws, and Breguet spring. Signed case, dial and movement.

REF. 6363 - ALBINO

$505,000 – NEW YORK, 23 APRIL 2008

Known and desired throughout the world, the Daytona Cosmograph reference 6263 is Rolex's sportswatch par excellance.

This reference is invariably associated with black dials with contrasting light subdials or silver dials with black subdials. The so-called 'Albino' dial, originating from Rolex's cutting-edge research in the 1970s, instead features a monochrome design, reminiscent of the dials on ref. 6238. The *maison* was seeking to lend a new aesthetic touch to a very well-known and widespread model. The white version of the dial, however, was never brought into production; it is believed to have been used for three copies alone, making them extremely valuable and sought-after. One of the famous 'Albino' Cosmographs was owned by the famous musician Eric Clapton. This watch was sold by Sotheby's in April 2008 for $505,000. Clapton is acknowledged to be a leading Rolex collector and has often been photographed with a watch on his wrist. Legend has it that Clapton has a rather extensive collection. The watch put on the auction block has a silver dial with a sunray finish, and the meticulous graphic detailing make it highly unlikely that this would have been a prototype.

Waterproof tonneau-shaped case with screw-down back, 36mm in diameter, screw-down crown and chronograph push-pieces. Ring calibrated for 200 units per hour. Matte silver dial, applied baton hour-markers, three subsidiary dials for constant seconds, 30-minute and 12-hour registers in guilloché silver. Calibre 727 movement with 17 rubies, mono-metallic self-compensating balance. Rolex Oyster steel bracelet with folding clasp. Signed case, dial and movement.

ROLEX
OYSTER
COSMOGRAPH
UNITS PER HOUR
SWISS

REF. 6284

CHF662,000 - GENEVA, 27 MARCH 2011

The starting price was CHF130-180,000, but as soon as the lot was presented, the price started rising – and it continued to rise until it reached CHF662,000: three times the initial sum.

This Oyster Perpetual reference 6284 is an extremely rare watch: it is one of the six with a *cloisonné* dial to have been manufactured by the *maison* circa 1950. In the 1950s Rolex worked together with the Geneva-based company Stern Frères Fabrique de Cadrans to produce a series of wristwatches with a *cloisonné* dial. These dials were extremely expensive to make, as they were hand-crafted by an expert artisan.
The dial on this model with a yellow-gold case depicts the two Americas. It was created by Marguerite Koch, one of Geneva's leading enamellers, who worked for the Stern company between the 1940s and '60s.
The artist created the outline of the motif by arranging slender gold threads on the base of the dial. The cells thus formed were filled with metal oxides in small quantities, which turned into enamel when the piece was placed in a furnace at around 900° C (1652° F). Finally, it was hand-polished to obtain a perfectly smooth surface. These valuable watches, produced in small series in yellow or pink gold, were all equipped with such dials - as in the case of ref. 6284.

Waterproof hand-wound watch with two-body case in yellow gold and screw-down back. The cloisonné enamel dial shows a map of the two Americas. Applied faceted yellow-gold hour-markers; dots on the burnished gold bezel indicate the minutes and seconds. Alpha hands. The watch runs on an automatic A.260 calibre

REF. 6538

CHF519,750 - GENEVA, 13 MAY 2013

At the Christie's auction held in May 2013, a new world record for the Rolex Submariner series was achieved by a reference 6538 watch, which was sold for 519,750 Swiss Francs.

Introduced in 1954, the Submariner was the first model designed for professional deep-sea exploration, as it allowed divers to measure their immersion time. The Submariner is also known as the James Bond model, as it is worn by Sean Connery in the films based on Ian Fleming's novels – the writer himself was a Rolex enthusiast.

The first references were waterproof to 100 metres, but then in 1957 reference 6538 was introduced, which is waterproof to 200 metres. This model is nicknamed "Coroncione" (Big Crown) because its crown is 8mm larger than those commonly used.

The watch sold at the auction is a rare 1959 exemplar with an enamelled black dial (also known as Patrizzi dial) with luminous hour-markers, which over time has acquired a brownish hue. The detail that adds most value to this Submariner is its 4-line dial with the text "660 ft = 200mt" and "Submariner", plus the words "Officially Certified Chronometer" – all at 6 o'clock.

Rolex Oyster Perpetual with tonneau-shaped, 38mm steel case with screw-down back and crown. Polished brown 4-line dial with dot and baton hour-markers and luminous numerals. Luminous Mercedes hands, small second hand at the centre. Rotatable black ring calibrated for 60 units. Extendible Oyster bracelet. Signed case, dial and movement.

50
10
20
30
200m = 660ft
CHRONOMETER

ROLEX GENEVE
17 RUBIS

THE TECHNOLOGY

This is a COSC-certified chronometer, one of the most accurate on the market.

GARANTIE Chronomètre

MONTRES ROLEX S.A. garantit la bonne marche de ce chronomètre pour la durée d'une année à partir de la date d'achat. Sont exceptés: la perte, le vol et les dommages résultant d'un traitement non approprié – y compris l'adjonction ou la substitution de pièces ou d'accessoires qui ne sont pas d'origine.

Nous certifions que ce chronomètre a été entièrement manufacturé en Suisse.

Nom et adresse de l'acheteur:
Date de l'achat:
Nom et adresse du bijoutier:

Cette garantie n'est valable que signée et datée par le concessionnaire officiel Rolex au moment de l'achat. Elle doit être remise avec la montre à examiner au concessionnaire qui l'a vendue ou à l'une des principales adresses de notre réseau mondial.

MONTRES ROLEX S.A.

ATTESTATION DE CHRONOMÈTRE OFFICIEL

Nous certifions que le mouvement de cette Rolex Oyster Perpetual a subi avec succès, pendant quinze jours et quinze nuits, en cinq positions et à trois températures déterminées, tous les contrôles de précision effectués par le Contrôle Officiel Suisse des Chronomètres (C.O.S.C.). Ce mouvement a ainsi acquis le droit de porter le titre de Chronomètre Suisse Officiellement Contrôlé.

A la suite de ces sévères épreuves, le mouvement et le boîtier ont encore dû passer un dernier et rigoureux examen dans nos propres laboratoires afin d'obtenir le droit de porter le titre de Chronomètre Rolex Oyster Perpetual à Sceau Rouge, titre attesté par les mots «Superlative Chronometer Officially Certified» qui paraissent sur le cadran et par le Sceau Rouge Rolex qui accompagne ce chronomètre.

MONTRES ROLEX S.A.

The technological development of the watches is based on three fundamental principles defined by Wilsdorf: accuracy, reliability, and practicality.

Accuracy

The first step towards accuracy was taken in 1910. Wilsdorf submitted a precision movement for analysis at the Bienne watchmaking school, later to become the *Bureaux Suisses de contrôle officiel de la marche des montres*. The very same year, this 11''' (24.81mm; 7/0) movement was the first to earn a wristwatch a Certificate of Chronometric Precision. In the wake of this success, an even smaller movement was presented, measuring just 9''' (20.3mm; 12/0). This movement prototype only earned the certificate in 1913.

On 15 July 1914, Rolex received a Class A Certificate of Chronometric Precision from the Kew Observatory in England: it was the first time this certificate was assigned to a wristwatch. For 45 days, this 11''' (24.81mm; 7/0) watch underwent five position tests and three temperature tests.

Technical testing had shown that the serial production of small high-precision movements was possible: this enabled Rolex to sell complete series of wrist chronometers with a Certificate of Chronometric Precision. In 1951, the Federation of the Swiss Watch Industry ruled that watchmakers could only describe their watches as chronometers if they had obtained the Certificate.

In 1924, two even smaller watches – a 73/4''' (17.483 mm)

TIMED
6 POSITIONS
OBSERVATORY QUALITY
ELINVAR
HAIRSPRING
F S
ROLEX
EXTRA PRIMA
SWISS MADE
15 RUBIES
70119

Beryllium balance with 18 screws, used for manual Cosmographs – a Rolex patent from 1960.

Opposite page
The first version of the Superbalance (top). The second version of the Superbalance (bottom).

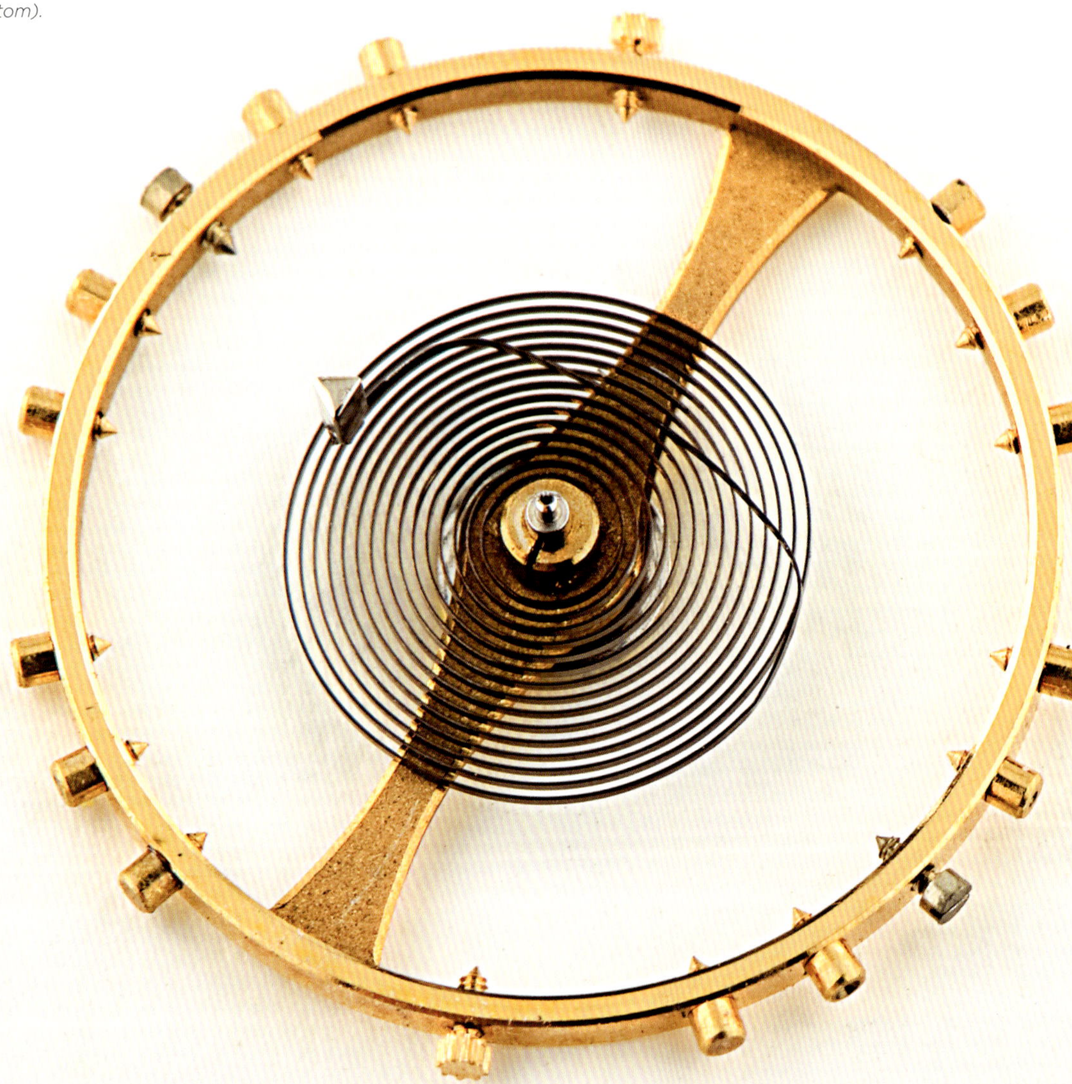

18 RUBIES
SWISS
MADE
ROLEX
PATENTED
SUPERBALANCE
E 83012

TIME
WITH
BALANCE
SCREWS
4162
PATENTED
SUPERBALANCE
KEW*A*TESTED
ROLEX
SWISS

round one and a 6''' (13.536mm; 20/0) oval one – obtained a Timing Certificate (of non-chronometric precision), again from the Kew Observatory.

By 1928 the number of certificates had doubled.

Rolex's research and development workshops designed a range of new balances unaffected by temperature: the precise Prima, Extra Prima and Ultra Prima, which could be adjusted in six positions and at all temperatures.

The Rolex archives show that in 1930 the *maison*'s watches earned 3,574 certificates, many with a mention of honour. Yet another record was set in 1935, when all of the 500 wristwatches presented received the certificate, in many cases with a mention of honour.

In 1941, Rolex started manufacturing watches with a new balance, one with even better performances and unaffected by the temperature: the "Superbalance". This balance was also used for the famous model that collectors nowadays refer as "Ovetto".

The movements were certified to have a maximum variance of ±7 seconds per day. On average, for every 100 pieces, the variance was of ±2.20 seconds (quality watches nowadays usually have a variance of ±3 seconds). Today, all Rolex watches are "Superlative Chronometer Officially Certified".

The four simple words engraved on the dial of a watch show that its movement has endured fifteen days and nights of testing, obtaining the COSC (Contrôle Officiel Suisse des Chronomètres) Certification.

The Parachrom hairspring

A watch is only ever as accurate as its balance: this functions as a flywheel running in both directions through the elastic force of the hairspring. Conventional hairsprings are made from a ferromagnetic alloy that is usually vulnerable to magnetic fields and shocks. In 2001, Rolex created the blue Parachrom hairspring. Made from a paramagnetic alloy, it is unaffected by magnetic fields and ten times more resistant to shock. The distinctive

The blue Parachrom hairspring is anti-magnetic and self-compensating. This watch is set with an inner Microstella balance.

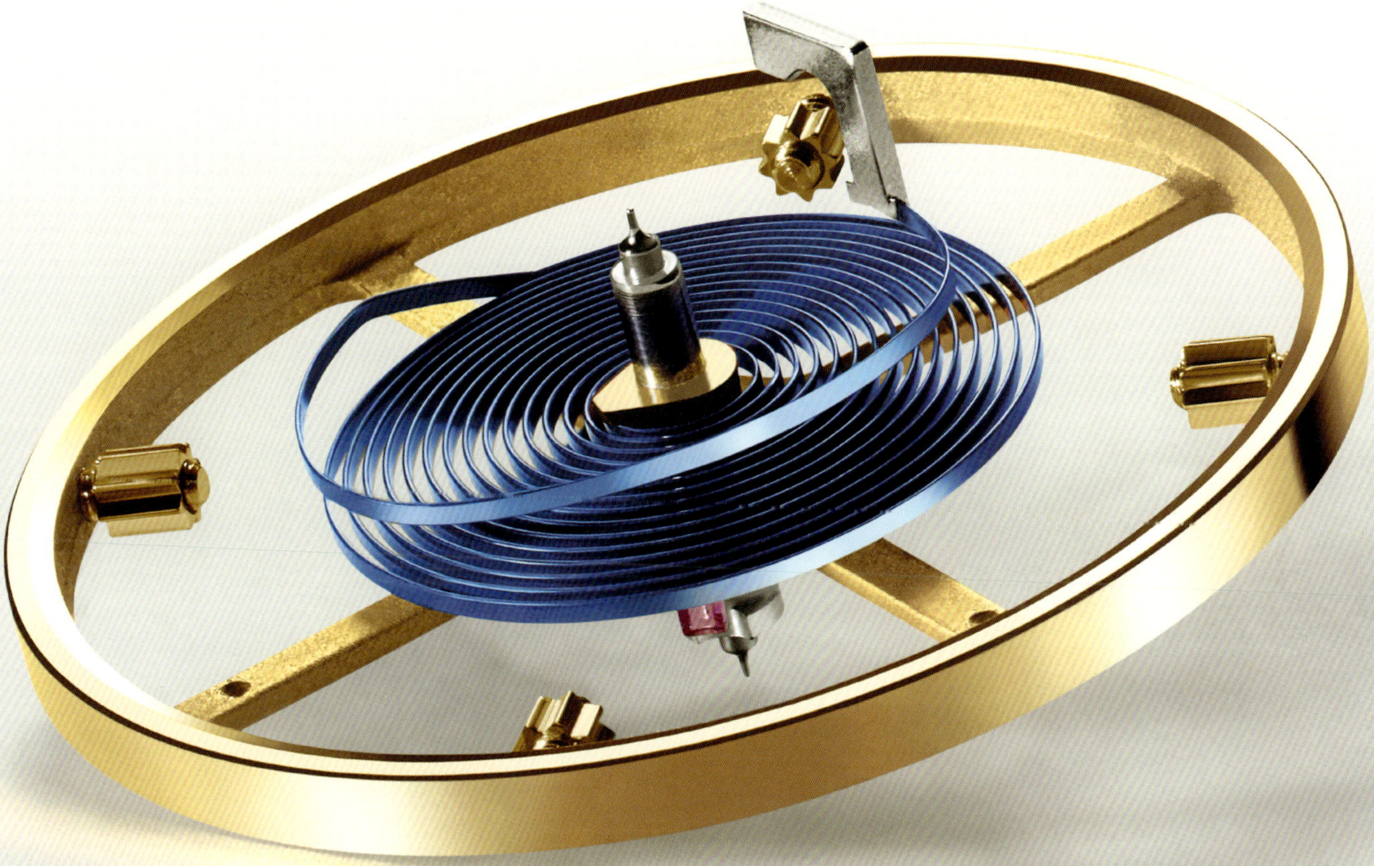

Rolex adverts illustrating the concept of waterproofing and the advantages provided by the rotor, which enables the watch to run smoothly and silently.

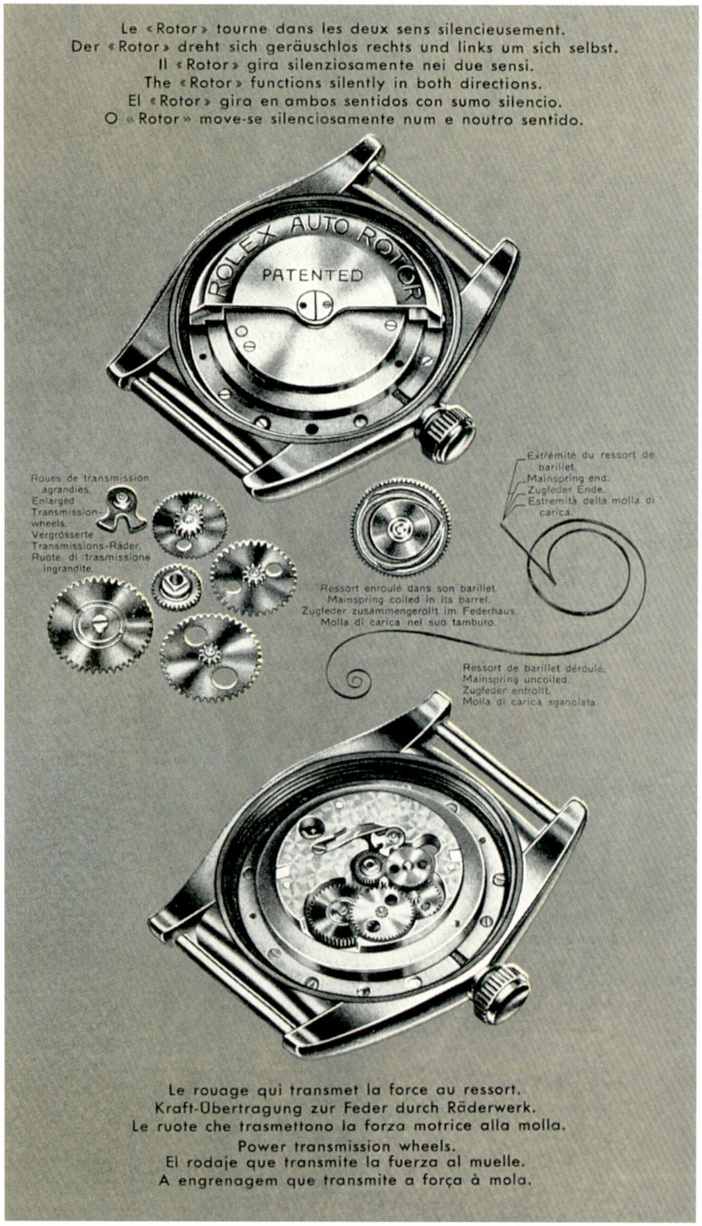

blue colour of the balance hairspring used to be a luxury feature of the most exclusive chronometers. Nowadays, a vast range of Oyster movements have the Parachrom hairspring.

Reliability

The second aim Wilsdorf had set himself was reliability. According to the vision of the founder of Rolex, reliability means providing a complete guarantee that, in any extreme situation, the timepiece will always be capable of providing the information required. The path towards this goal was a long and difficult one. A wristwatch must be protected from shocks, dampness, sweat and dust. If we add the fact that for most of the day a person will be constantly moving, and if we multiply this movement for days, months and years, we realise just what kind of difficulties Rolex engineers had to grapple with.

An initial difficulty concerned which metal was to be used and how it could be worked. A second difficulty was to ensure the complete waterproofing of the case, crown, and crystal. Wilsdorf came up with a solution to the problem of making the crown waterproof in 1926, when he purchased the patent for the screw-down top from its two inventors.

This system became even more effective a few years later with the development of the self-winding movement. Having to charge one's watch every day, by screwing and unscrewing the crown, would wear and harden the rubber gasket inside, increasing the risk of seepage. This carefully developed and streamlined crown is now known as the Triplock or Twinlock, depending on the model, and is mounted on all waterproof watches.

In 1926, Rolex invented and patented the case with a screw-down base, ensuring even greater waterproofing.

The Triplock winding-crown and the screw tube are equipped with five gaskets to ensure waterproofing.

Making the most of its latest discovery, Rolex launched the first fully waterproof watch, which was registered under the name of Oyster.

Reliability also means protecting the watch from shocks: in 1955, the Rolex Shock Resisting anti-shock system was adopted.

In order to protect its watches, Rolex had developed an anti-shock system called Paraflex. The Paraflex spring increases shock resistance by over 50%, and with it the reliability and durability of the movement itself.

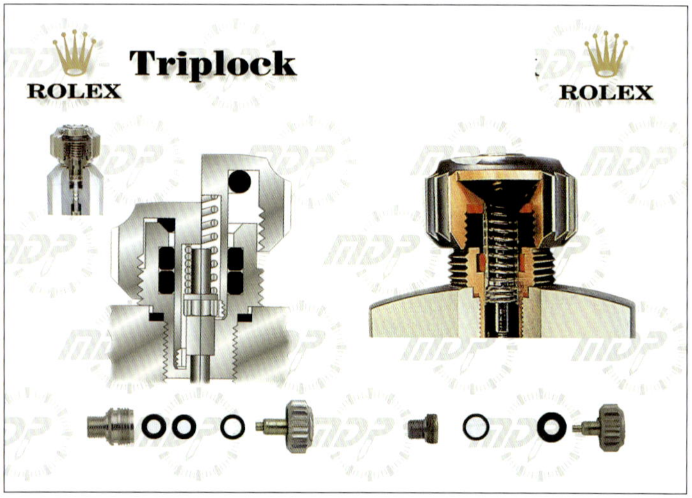

Access denied

In order to protect the movement from water and dust, Rolex invented and patented the Triplock screw-down winding crown. The professional Submariner and Rolex Deepsea models equipped with this ingenious triple-sealing system are waterproof to 300 and 3,900 metres, respectively.

The Triplock crown consists of ten components made of precious metal alloys – such as 18ct gold or 904L steel – and special polymers for the gaskets; once assembled, the crown is screwed onto the Oyster case like a submarine hatch. As a guarantee of Rolex quality, the pressure levels tested are much higher than those at the depth for which waterproofing is certi-

In 1931, the Perpetual movement was developed with a winding rotor. Set at the centre of the movement, this rotates within a hollow space around the mechanism and is free to move at 360°, preventing any loosening of the screws. The rotor is separated from the rest of the mechanism by a cap. The rotor patented by Rolex rotates in both directions, transmitting a continuous and regular driving force.

fied – a depth to which few underwater devices have ventured so far.

Practicality

The third necessity laid down by Wilsdorf is practicality. Practicality means that a watch must not only be beautiful, reliable and accurate, but also easy to use.

The most striking, practical and revolutionary invention made by Wilsdorf and his team of engineers is no doubt the self-winding mechanism known and patented under the name of Perpetual (which refers to perpetual winding rather than the perpetual calendar function). The movement of the wearer's wrist naturally creates the energy required for winding the watch.

The self-winding calibre 4130 movement of the Rolex Daytona, in use since 1988.

Top right
The Parashock Ultraflex system.

THE MATERIALS

904L steel

For its watch cases, Rolex uses 904L steel, which is mostly employed in the chemical industry to ensure maximum resistance to corrosion. This extremely resistant and shiny alloy is combined with a range of other precious materials.

Everose gold

In searching for a colour that would not change over time, Rolex created the gold alloy Everose. Pink gold tends to lose its original colour because of external agents – especially the chlorine in water. The platinum in the Everose alloy protects it from harmful elements and fixes its colour for good. Tests have confirmed that the Everose alloy will endure over time without any alteration.

Cerachrom

The graduated bezel of a watch may be damaged by sunlight or the chlorine in water, or may become scratched. To guard against these problems, Rolex uses a bezel with a Cerachrom ring, made of a very resistant ceramic with outstanding anti-corrosion properties, making it colour-fast and scratch-resistant.
To engrave numerals and hour-markers, Rolex has developed and patented a special procedure that is applied before the hardening of the ceramic. The disc is then fully plated with yellow or platinum gold, applied atom by atom, and is then smoothed down so that the precious metal will only remain in the numerals and hour-markers. It takes 40 hours to manufacture a bezel of this kind.

GLOSSARY

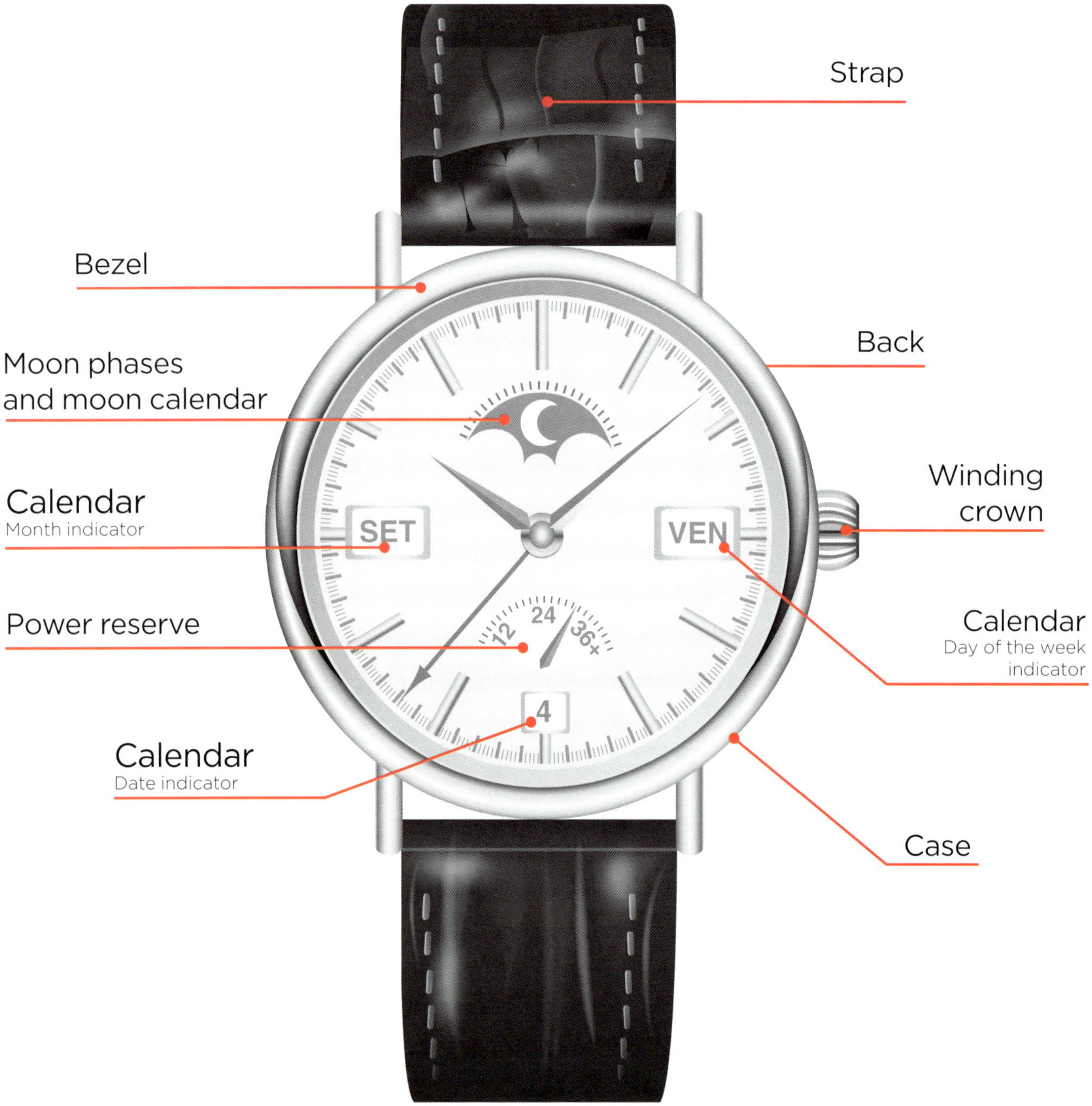

Strap
Bezel
Back
Moon phases
and moon calendar
Winding
crown
Calendar
Month indicator
SET
VEN
24
12
36+
4
Power reserve
Calendar
Day of the week
indicator
Calendar
Date indicator
Case

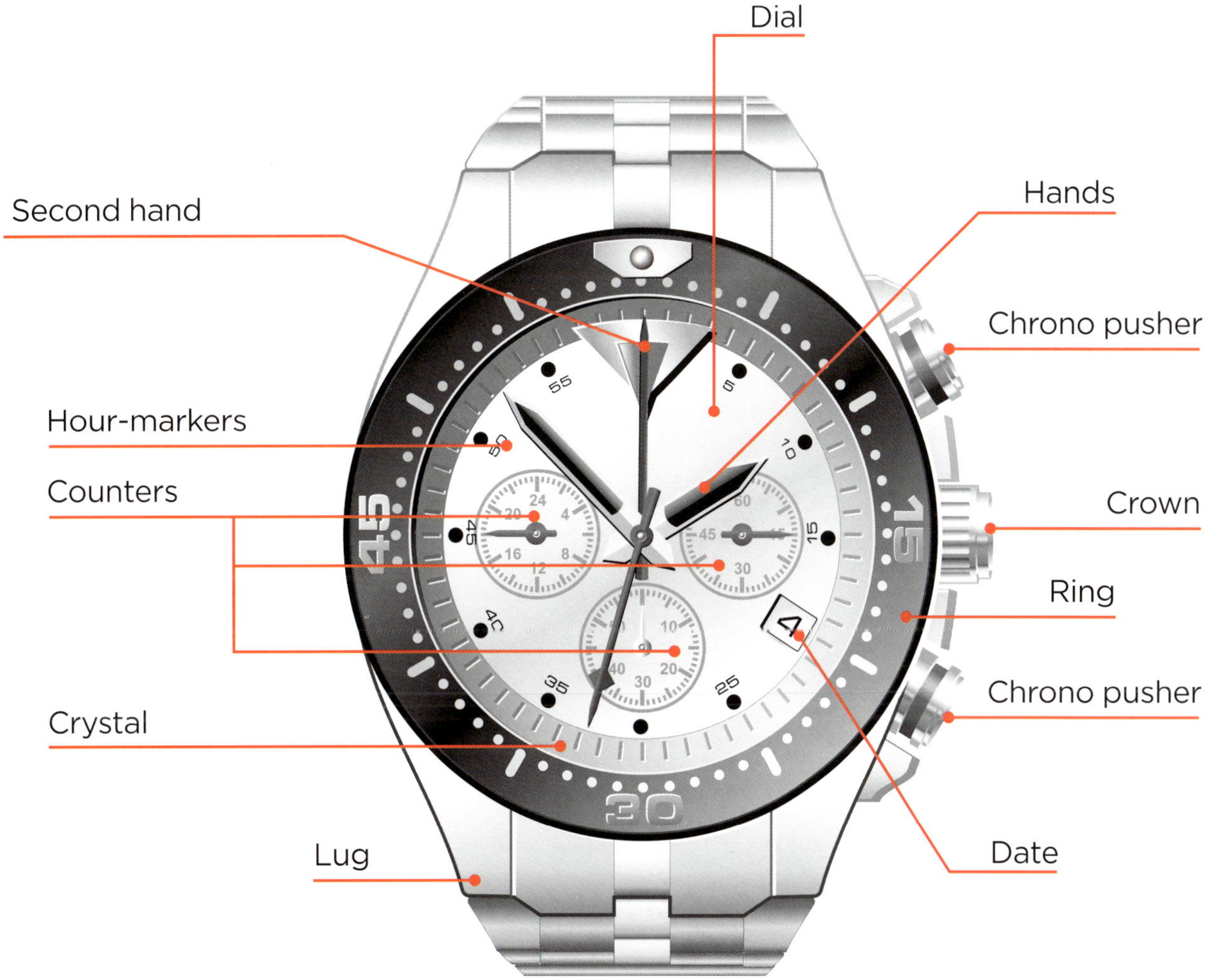
Dial
Second hand
Hands
Chrono pusher
Hour-markers
Counters
Crown
Ring
Chrono pusher
Crystal
Lug
Date

Alarm wristwatch: a watch with a mechanical alarm added to the usual hour, minute and second indication. Generally, the crown autonomously controls the setting of the time, the winding of the specific spring and the setting of the alarm.

Back: rear part of the case. It can be removed to access the movement. It is either pressed into position or secured with screws.

Back with visible movement: usually found in automatic watches or high-end mechanical watches. Instead of a metal base touching the wrist, there is a sapphire or mineral glass back allowing a clear vision of the movement and the inner workings of the watch.

Balance: pivoting on the movement, it serves as a fly, moving in two directions thanks to the elastic force of the spring.

Barrel: the core of the movement. It contains the mainspring, which serves as first wheel of the going train (the series of wheels and gears that form the movement) for the transmission of the motive force.

Baseplate: the "framework" of the movement. It offers support to all the mechanical components assuring structural strength to the whole system. Disc-shaped, the baseplate is commonly made of brass and is divided into sections where all the different components are collocated.

Bezel: usually the upper part of the watch body. In particular, it usually refers to a ring set around the external part of the crystal. In the jewel watches, for example, the bezel can accommodate a ring of diamonds. In sports watches, the bezel can rotate in one or two directions and have some graduated indications. In this case it is called a ring.

Bridges: along with the baseplate, they form the "sandwich" structure supporting the wheels and the different movement components. They are made up of distinct parts (barrel bridge, train wheel bridge, balance bridge etc.) to make disassembly of the movement easier for the watchmaker.

Cabochon: in gold jewellery it indicates an oval cut with no faces. It is used for wristwatch crowns with decorative purposes.

Calendars:
Complete calendar: indicates the date, the day of the week, the month and often even the moon phases. Like the "simple calendar", it must be manually adjusted at the end of any month with fewer than 31 days.
Perpetual calendar: self-regulating calendar that automatically adjusts to the different durations of the months (28, 30 or 31 days) and leap years.
Simple calendar: indicates the date and sometimes the day of the week. It requires manual adjustment at the end of any month with fewer than 31 days.
Yearly calendar: indicates the day, the date, the month, the 24 hours, with an automatic regulation for the long and short months. It needs adjustment just once a year: at the end of February.

Calibre: this term originally indicated and defined the diameter of a movement. Today it identifies the different types of movement with a specific brand and number.

Carrure: central part of the watchcase where the movement is set. The lugs depart from the *carrure*. The back and the bezel supporting the crystal are fixed onto the *carrure*. It can be smooth or ribbed (*cannelée*).

Case: unit comprising the bezel, *carrure* and back. Produced in different shapes and materials, it contains the movement of the watch.

Chronograph: thanks to the push-pieces and the special internal mechanism, the chronograph can measure precise time intervals. By means of a play of levers and wheels, when the push-piece start is pressed, a series of wheels (onto which the chronograph second, minute and hours are fitted) hook onto the usual second wheel and are set into motion. To stop the mechanism, the clutch wheel is distanced from the normal second hand. Usually, you start the chronograph with the push-piece at 2 and you reset it at 4. There are, however, old mechanical chronographs where the stop and the reset are activated by pressing the push-piece at 4.

Chronometer: a highly precise watch whose movement, during a series of checks, produces a time range between -4 and +6 seconds a day.

Complication: any addition to the simple time indication function on the dial (hours, minutes and seconds) of a hand-wound movement. Although being very common, both the date and automatic winding system are considered "complications".

COSC: acronym for "Contrôle Officiel Suisse des Chronomètres", the Swiss institute that isues the certificates guaranteeing the precision of wristwatches. A COSC watch can be identified by a serial number inscribed on the movement.

Counter: a complication; a small and usually round part of the dial.

Crown: an element mounted on the *carrure* to hand-wind the watch. When extracted, it regulates the hands and the date.

Deployment clasp: a fastening system divided in three parts folding onto each other. When opened, the two extremities of the watchband remain fixed and let the hand pass through. When closed, the buckle clicks onto the two pieces that fold onto each other.

Double time-zone display: a watch that measures local time and at least one other time zone. This function can be indicated on a twin dial, with an additional hand, a secondary-dial, or with other solutions. Another name for this is "world time function".

Escapement: the mechanism composed of the escapement wheel and pallet that on one side transmits to the balance the impulse necessary for the upkeep of the vibrations and on the other side "manages" the time-telling process by letting the escapement wheel teeth "escape" one at a time, at each vibration of the balance.

Frequency: it indicates the per hour cadence of the balance. 18.000 or 21.600 per hour cadences are common in movements produced until the 1960s, while 28.800 or 36.000 are found in more recent movements.

Gmt: watches that display on the dial a second world time with a 24-hour hand and a reference ring with 24 marks.

Hairspring: one of the most important and delicate elements of the movement. The spring that makes the balance oscillate by setting the "pace". It is responsible for the balance frequency (the maximum rotation of the balance in the two directions = 1 oscillation or 2 movements).

Index: a marker on the dial indicating the time or a function.

Jump-hour: a watch that does not tell the time with the usual hand system, but with a digital indication (a number) inside an opening. The name derives from the fact that at the stroke of the 60th minute the indication "jumps" to the next number, just like a date marker.

Lug: the part connecting the *carrure* of the wristwatch case to the watchband.

Minute repeater: a watch that can be set to strike the hours, the quarters and the minutes with three different gong pitches, for an acoustic signalling of the current time. The function is activated by moving the repeater slide on the case or by pressing a push-piece.

Moon-phase: indication of the moon-phases. The moon carries out a regular rotation around the earth every 29 days, 12 hours and 44 minutes. Once set, the indication shows precisely the current moon phase.

Power reserve: indication signalling the level of power still available. It appears on the dial (usually with a hand on a circular section, but also with a small window) and it indicates the hours left until the end of the power reserve.

Power reserve indicator: indicates the remaining autonomy in the movement, usually expressed in days. A hand on the dial describes a circle or a sector and is connected to the barrel with a module calculating the power reserve still available in the spring.

Push-pieces: small protruding buttons to start, stop or reset the chronograph.

Retrograde indication: a hand that instead of offering an indication running along a 360° rotation (such as hours, minutes, date, etc.) carries out only a portion of the circle and then starts a new measurement after returning to the starting point.

Scales: usually positioned along the outer flange of a chronograph dial, they measure speed (tachometer scale), respiratory rate (asthmometer scale), pulse rate (pulsometer scale), and the distance in km from a luminous event followed by a sound, such as a flash of lightning followed by thunder (telemetric scale).

Screw-down crown: its purpose is to press with a certain force the rubber or synthetic sealing that assures protection at the level of the stem.

Spilt-seconds chronograph: a special type of chronograph that thanks to an additional mechanism and chronograph hand can track intermediate times (one hand stops on an intermediate time, while the other continues its course), or two different times (one hand stops before the other).

Tourbillon: patented in 1801 by Abraham-Louis Breguet, the concept of the *tourbillon* is to insert the regulating mechanism of the watch into a very light cage that generally rotates around itself, once every minute. This delivers a very regular movement and extreme precision in the marking of time. The *tourbillon* mechanism tends to reduce the irregularities in the movement of the watch due to gravity.

Two-directional ring: bezel that can move clockwise and anti-clockwise; usually employed for calculations or time measurements.

Unidirectional ring: to mark the maximum time of immersion (to mark the duration of a scuba tank, for example). It can only turn in one direction to prevent the indicated time being accidentally increased.

Watch-jewels: bearings mounted on the bridges and the baseplate in which the pinions rotate. Their purpose is to retain the lubricant oils to minimise friction and wear.

Winding spring: fitted inside the barrel, it constitutes the engine of the entire system. With either a hand-wound or automatic charge, this spring stores up the energy needed for the functioning of the movement.

BIBLIOGRAPHY

Mondani, F. and Mondani, G., *100 years of Rolex 1908-2008*, (English/Italian dual language), Genoa, Guido Mondani Editore, 2008.

Patrizzi, O., *Collecting Modern and Vintage ROLEX Wristwatches Vol I* (English/Italian dual language), Genoa, Guido Mondani Editore, 2006.

— *Collecting Modern and Vintage ROLEX Wristwatches Vol II* (English/Italian dual language), Genoa, Guido Mondani Editore, 2006.

— *Cosmograph Daytona Manual Winding*, (English/Italian dual language), Milan, Action Group Editore, 2010.

— *Oyster Perpetual Cosmograph Daytona*, (English/Italian dual language), Geneva, Patrizzi & Co. Edition, 2011.

PHOTOGRAPHIC CREDITS

For all images, if not otherwise specified © Archivio Guido Mondani Editore, Genoa

Additional photographic references:
© Christie's Images Limited 2014: 102-121, 124-125, 130-131
Courtesy Antiquorum Auctioneers: 123, 128-129
Courtesy John Goldberger: 21 (left), 32-33, 135, 137
Courtesy Patrizzi & Co. Edition: 5 (above, middle and below right), 6-11, 25, 44-45, 52-53, 70, 73, 76, 84-87, 89, 94-95, 97, 100-101, 132-133, 136, 142, 144
Courtesy Sotheby's: 127
© Shutterstock Images LLC: 146-147.

The authors and publisher gratefully acknowledge the permission granted to reproduce the copyright material in this book. Every effort has been made to trace copyright holders and to obtain their permission for the use of copyright material. The publisher apologises for any errors or omissions in the text and would be grateful if notified of any corrections that should be incorporated in future reprints or editions of this book.

ISBN: 978 185149 783 6

Published in Italian under the title:
Rolex. La storia, le icone e i modelli da record

ISBN: 978-88-6648-199-7

British Library Cataloguing-in-Publication Data

A catalogue record for this book is available from the British Library

Printed in China
for ACC Art Books Ltd., Woodbridge, Suffolk, England

SWISS